BIODYNAMIC WINE,
DEMYSTIFIED

NICOLAS JOLY

Wine Appreciation Guild
San Francisco

Biodynamic Wine, Demystified

Published by
Wine Appreciation Guild
360 Swift Avenue
South San Francisco, CA 94080
(650) 866-3020

www.wineappreciation.com

Library of Congress Cataloging-in-Publication Data

Joly, Nicolas.
Biodynamic wine, demystified / Nicolas Joly.
p. cm.
Includes bibliographical references and index.

ISBN 978-1-934259-02-3

1. Viticulture. 2. Wine and wine making. 3. Organic farming. I. Title.
SB388.J65 2007
634.8--dc22

2007035464

Contents

Prologue

Observations we can prove we call science. Those we cannot prove we call religion. Rigorous scientists and religious fundamentalists consider most anything else superstition, which is a label often applied to biodynamics.

Nicolas Joly, one of the foremost proponents of biodynamic viticulture, has a talent for observation. Through his own personal study of the forces acting on his vines at one of the world's great vineyards, the Coulée de Serrant in Savennières, Joly has looked beyond material science to nurture and strengthen his vines and the environment that surrounds them. While most commercial agriculture manages plants and animals as a means of production, Joly and a growing number of his contemporaries have sought an approach that respects and responds to plants and animals as living, sentient creatures. These biodynamic farmers take their cue from a series of lectures given by Rudolf Steiner in 1925. In them, Steiner offered observations of traditional farming practices; he took a respectful view of human interaction with forces driven by the earth and the sun, the moon and stars.

To some, such a view immediately sounds romantic or foolish, as if science had at some point invalidated these observations. In fact, a number of these observations and practices show results that can be measured scientifically; others await further study.

Joly and many of his talented companions in biodynamics are neither romantic nor foolish. They have interpreted Steiner's observations and applied them to viticulture with the goal of making the most distinctive and authentic wine from their vineyards. Their farming not only requires more rigorous care than most, it also requires personal decisions based on constant observations of the life around them and the forces acting on that life, whether it's the invisible life of microbes in the soil or the position of the sun.

Their work has refined the expression of some of the greatest wines in the world.

Readers who come to this book looking for a set of practices to follow at their own vineyard, farm or garden will be disappointed.

Those who are looking for evidence to prove the effectiveness of biodynamic practices will have to turn to some of the works cited in the bibliography at the end of this book. The philosophy that drives biodynamics remains radical enough to require its own carefully articulated statement. This book is not only a consideration of what is authentic in wine. It is a search for a man or woman's place in this world, as neither a scientist nor as a priest, but as a farmer. Joly has written an essay on what it means to be human, and to be humane, in the best sense of the word.

Joshua Greene

Foreword

If a vineyard had Nicolas Joly's energy, character and depth, it would be a first growth. As a noble protector of Gaia and a tireless advocate for her healing through the practices of Biodynamics, Joly and his writings offer hope for the health of our planet in a world seeking answers.

With groundbreaking insight, Joly's first book – *Wine from Sky to Earth* – addressed the confusion and esoterism that have discouraged some growers and winemakers from embracing Biodynamics, while clarifying many of the mysteries that historically have challenged its legitimacy as a farming practice for wine. As the owner/winegrower of the first certified-Biodynamic vineyard estate in Sonoma County, California, I was not only inspired by the penetrating observations in this book to deepen my commitment to Biodynamics, but also empowered to better explain its complexities to my family, neighbors and customers.

In his new book, *Biodynamic Wine, Demystified*, Joly once again uses vivid metaphors and astute comparisons to blow the subject of Biodynamics wide open without watering it down. This time, he is writing for an audience that includes the wine enthusiast new to Biodynamics. With more and more people concerned about how the products they buy are made, and how these products impact the environment and personal health, this book arrives in an hour of need.

Throughout the book, Joly urges a powerful shift in thinking: To begin to understand the physical, living world, we must consider the forces and energies that give it shape. As Goethe wrote, "*Matter is nothing, what counts is the gesture that made it.*" I share the belief that in working with formative energies respectfully and at the right time in the right way, we can create a healthier environment and wines that capture the flavors and aromas of a particular place and time.

Biodynamics is, at its core, an energy management system. When practiced rightly, it brings a dynamic balance to the land, enabling the winegrower to realize the maximum potential for that vintage. This is because a vine tended under these conditions becomes more than a plant responding to stimuli, it becomes a super-sensitive life form with the ability to order and organize energies that manifest themselves as varietal character, place, vintage and even intentionality.

What can Biodynamics do for our environment and the world of wine? *Biodynamic Wine, Demystified* provides a thought-provoking response. Perhaps better than anyone, Joly understands that the foundation of Biodynamics is the personal relationship a farmer has with the land. Over time, Biodynamics allows the farmer to see the land with new eyes. As Marcel Proust noted, "*The real voyage of discovery consists not in seeing new lands, but in seeing with new eyes.*" This is the new consciousness needed to heal our planet. As support for Biodynamics grows with the help of visionaries like Nicolas Joly, the impact of this most healing form of agriculture will resonate around the globe.

Mike Benziger, Benziger Family Winery

Glen Ellen, California

October 2007

Preface

The sole aim of this book is to forge a link between a knowledge existing since the dawn of time that is profound and endlessly available – but not understood – and a science which, while it knows almost everything, nevertheless understands next to nothing.

Nicolas Joly

1

Passion for Wine and the Appellation d'Origine Contrôlée (AOC)

'Matter as such does not exist; all matter originates, and exists, solely by virtue of a force which induces particles to vibrate.' Max Planck , physicist, Nobel Prize acceptance speech.

A passion for wine is spreading like wildfire through the world, like a quest for something to give life greater sense and joy. Wine-lovers compare the idiosyncratic tastes of grapes growing in different locations of the globe. With something bordering on apprehension they relish the brand and trade name implacably imposed on a grower despite an increasingly disrupted climate. They leave the bottle open to 're-taste' it the next day and the day after that. They deliberate, calculate, wonder, question and get carried away with their particular enthusiasms. This is indeed a growing passion which touches all professions and social classes, which sharpens each person's senses, impressions and emotions. Though some degree of knowledge develops from this passion, it never quite hardens into certainty. It remains, instead, something against which to continually test one's faculties and one's desire to apprehend the realities of another, fluctuating and intangible world – that of aromas, tastes, balances and harmonies.

A fragile world, for which those of an artistic sensibility always feel a certain nostalgia, which expresses itself subtly, discretely and almost shyly through matter. We desire to understand how an equilibrium sometimes so delicate is achieved; how these bright and dark moods, these sorrows and joys of the vine can ultimately become tastes, scents or harmonies of an almost musical nature.

Basically all this underlies, and justifies, the profound concept of the 'appellations contrôlées' or 'regulated wine of origin.'[1] Back in the 1930s when France, followed swiftly by many other countries, created the AOC standards, what was its aim? It simply wished to protect a sum of knowledge, an accumulation of experience, a finger-tip feeling several centuries old that had led people to plant wine in certain 'good' locations. What did a 'good location' mean in that less abstract era? Quite simply a place where 'Lady Vine' felt at ease could give full 'voice' to her happiness and sing forth without hindrance. We will find that this song is not always as joyous as we think. For the moment it is enough to understand that, when a vine is situated where it can unfold its full potency as a highly atypical and self-willed vegetative being, it will imbue its fruit with a taste endowed by the place in which it grows. Simple enough? It weds the soil via its roots, uniting with

---

[1] 'Appellation Contrôlée' is a guarantee that a wine has been produced in a specific location (appellation), by a particular method, with approved grape varieties and in controlled quantities. The system is legally defined and regulated in France.

it intimately, and receiving through its leaves all the climatic conditions specific to that area. These are composed of the different qualities of heat which arise at different moments, of variations of light intensity, of winds full of gentleness or revolt, of modest or abundant rains, of morning mists or brief twilights: all these aspects of weather combine to become first vegetative matter and ultimately fruit. But how does this actually happen?

Take a look at a field of vines, in spring first of all, then in the autumn: you have to realize that all these branches, these leaves, and several tonnes of grapes per hectare – which were mere buds 6 months before – are barely composed of the substance of the soil, as people too often assume. On the contrary, the major part of their substance comes from photosynthesis, a wretched word shorn of beauty which does not come close to expressing a still unexplained mystery that the scientific world observes without being able to reproduce. Photosynthesis refers to the conversion of heat, light and air – a world, therefore, of almost intangible forces like the tastes and aromas we mentioned above – into real matter composed of carbohydrate, starch, sugars etc. If one excludes water from these substances – thus remaining with 'dry matter' as science terms it – over 92 percent derives from photosynthesis and thus only a very small amount can be attributed to the soil itself. From spring to autumn, too often without realizing it, we witness the plant world 'materializing' an almost invisible world, a process in which the agency of climate plays an important part. Into matter and substance descend subtleties of taste, color and scent so prized by wine-lovers: truffle, olive oil, coffee, cigar, tea etc. Each plant accomplishes this task in its 'own' manner, with its unique nuances which give us such pleasure if we know how to recognize them, and can distinguish them from the artificial flavors that technology secretly infiltrates into our food and drink.

With something akin to hypersensitivity the vine excels in its capacity to create nuances of taste. It is therefore interesting to try to understand in detail the deep nature of our friend the vine or, let us say, to enter into its secret gestures so as to approach the very nature of wine.

What place does the vine occupy in the plant kingdom? What is its character, its conduct, its unique nature? Like all living beings, none of whom are merely driven by blind cause and effect, this question takes us in an important direction. To answer it we need to return to the botanists of the Middle Ages and their rich store of knowledge, so little understood by our modern era. They had a very different view of plants from us. Matter itself, which we are so interested in nowadays, right down to its tiniest atoms, was for them merely something that served to fill a form, like the dough in a bread tin. What medieval scholars were interested in was the mould or form itself, in other words the various forces which 'sculpt' the vegetable world differently in each instance, and which give it a particular aspect and mode of behavior. This was nothing to do with genes – which of course they had never heard of. But if one had talked to them about genes they would probably have replied: 'Why concern yourself with the obedient laborers who merely carry out orders? Instead study the architects who arrange and organize these genes.' Thus they would direct us to the whole system of energies which physicists are just beginning to comprehend today through magnetic resonance imaging, something which biodynamics makes full use of. Reading Hildegard von Bingen[2], Culpeper[3] and many other famous authors of this period, an era so poorly understood by modern science, we find that all of them approach the plant world through what Plato calls the 'four states of matter' (see Plate 1). Thanks to this formidable body of knowledge one can develop a quite different perspective on the vine and wine.

This ancient wisdom can be briefly and simply, though very imperfectly, summed up as follows. The earth is subject to a force – gravity – that holds sway over every living being and thus also ourselves. It is gravity, this omnipresent force, which makes a stone fall when we throw it, which makes rain fall to earth, and which leaves us feeling so heavy after a day spent working hard. It is by virtue of this force that atoms coalesce, that matter forms and can attain a state of solidity. Without it the physical world, the earth's physical substance, would not exist.

---

[2] Blessed Hildegard of Bingen (1098–1179), a German teacher, monastic leader, mystic, author, and composer of music.
[3] Nicolas Culpeper (1616–1654), an English botanist, herbalist, physician, and astrologer.

Most fortunately, though, this force is counterbalanced by another, an opposite polarity. In the West we refer to this as 'solar attraction,' and in the East it is often described as the force of levitation. Acting in opposition to gravity this leads towards a state of weightlessness. In physical terms heat embodies this force most clearly, which is thus one of rising or lifting from the earth. Just observe how every flame emits an ascending shimmer of heat. Heat dispels and disperses matter. Heated up, a heavy piece of metal turns to liquid, and then soon enters a gaseous state, delivered of its weight. This reveals the impermanence of matter and the physical world, which oscillates between the visible and the invisible – a theme we will return to later (see diagram on opposite page). The human being is also subject to this force, and it is this which indirectly – a subject in itself – enables him to wake up in the morning feeling light and renewed. It is this, likewise, which lends us wings to soar above the day's vicissitudes when we hear a piece of good news, and which plays such an important part in feelings of enthusiasm.

The great sages of the past stated that there were two intermediary states between these two forces. Descending from above, from a more rarefied condition, the first of these is air and light. This is the first condition with a slightly terrestrial or physical quality. Air and light are closely connected, the latter becoming visible to us by means of the former. Without air the sky would not appear blue to us. Passing beyond the layers of atmosphere we find the sky is dark, opaque. The air has little weight but it is still, nevertheless, subject to gravity, a fact which, fortunately

Different states of a same substance through fire forces.

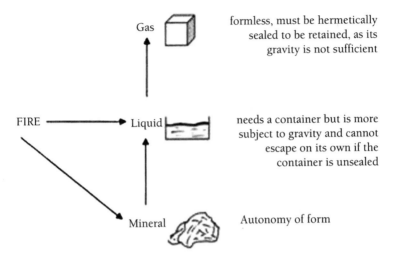

Gas — formless, must be hermetically sealed to be retained, as its gravity is not sufficient

FIRE ⟶ Liquid — needs a container but is more subject to gravity and cannot escape on its own if the container is unsealed

Mineral — Autonomy of form

(Source: Rudolf Steiner, The Warmth Course, Mercury Press 1988)

Fire/gas

for us, keeps it closely wrapped around and enfolding the globe. When compressed (excessive gravity) the air actually grows denser. These examples allow us to grasp the true nature of air and light as a first state of matter subject to density.

Next comes the liquid state. This condition, whose archetype is water, can be seen as occupying an intermediary position between the solar, ascending laws, and those of the earth. Water is more subject to gravity than air, and thus heavier and more earthly in nature. It is poised midway between the earth's gravitational attraction and the rising solar forces. It grows hard as a stone in cold conditions, clearly subject to gravity in a mineral-like fashion. Heat, on the other hand, releases it from earthly laws, enabling it to escape upwards as mist and fog. Archimedes tells us that water relieves us of some of our weight. Some of the pleasure of swimming is in being partly cushioned from the tug of gravity. We ourselves are composed of more than 90 percent water, which also helps explain the effect of various water treatments and therapies. As we will see, water is the bearer of resonance information, by means of which we stay alive.

In all this the important thing to note is that air and liquid are intermediary states in a progressive descent towards our earth's solid mineral substance. No life is possible on earth without passing initially through a liquid state – and the same applies of course to plants. It is important to remember also that a living organism arises through the ongoing interaction of these four states, often with one or another playing a dominant role.

But why this long preamble in a book about wine? Well, so as to develop some understanding of the way in which plants relate to these four terrestrial conditions, and thus discover the profoundly atypical nature of the vine.

One can say, in general, that each plant reveals its relationship to earthly forces of gravity in its roots, to the watery state in its leaves, to the light in its capacity to bloom, and to warmth and heat in its power to fruit (see diagram on opposite page). It should be added that these four aspects also interpenetrate each other in the plant.

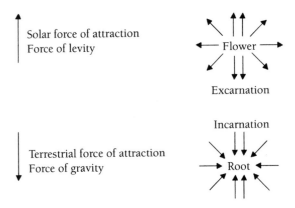

The four states of matter in the plant

Discovering the individual quality of every plant

All this could be described in length and detail, and should be understood in very specific, tangible terms. But where it becomes still more interesting is in the discovery that certain plants establish such relationships with *extreme originality*. Let's take a few simple examples to start with: a plant whose archetypal nature is to develop a strong affinity with the light will absorb it better, and can express or manifest this particular quality in a ravishing blossom. The lily is an example of this, a flower that adorned the flag of the French royalty. If the lily did not have this strong connection with light its power to bloom would be considerably less.

A different plant, closely connected with water such as rhubarb, will be able to produce big and abundant leaves. One intimately related to warmth and heat will receive from it a powerful fruit or seed-forming force. We can see this in grains and cereals where each grain sown multiplies a hundredfold.

All this becomes more complex when Mother Nature departs from her habitual schemas and plays with these four states of matter with such ingenuity and inventiveness that the result is sometimes very difficult to decipher. One example is the carrot which uses its floral capacity to color and perfume its root, thus rendering the flower itself a poor and unattractive specimen. The pine tree pours its strong connection with the forces of warmth not into fruit but into its inflammable resin, thus endowing itself with the capacity to resist a greater degree of cold than most other plants. The willow is strongly connected not just to water but also to light. Because of this it does not form big leaves but manages to exhale this water through its leaves, and to evaporate it at the same time as attaining a honeyed state in its delicate flowers, of which the bees are so fond. It can also color its wood a vivid yellow. The nettle (see Plate 3) does not put its connection with warmth into its seeds but into its leaves. Used as an infusion sprayed on vines, this gives them the capacity to maintain their sap circulation even at times of drought. Citronella takes the strength of taste destined for the fruit it does not form and puts it into its leaves. In the cinnamon, taste descends right into the bark! The toughness of wood can ascend somewhat into the leaf of the magnolia ... And so on, and so forth.

Thus there is a secret language sometimes very subtle and difficult to decode. We can admire the ancient scholars who first deciphered this script and then linked it on occasion to very specific planetary or stellar forces whose effect they identified in matter. They also used this knowledge in very precise ways to create powerful remedies. This ability to observe life directly can still be found amongst peoples said to be 'primitive,' who know of remarkable properties in the vegetable and animal world that surrounds them. The ancients did not have the same type of intelligence as we understand the word today, nor sophisticated instruments which our modern researchers depend on for their narrowly focused and strictly physical understanding of the world. They had other faculties instead which enabled them to tap into the deep messages of their senses. One can see this in very simple things such as the way people of the past knew how to cut a plant or tree at the right moment, in accordance with its native characteristics, or, let us say, its originating force. This optimized its effects somewhat like a musical note which uses its surroundings to resonate effectively. In olden times, for example, without any calendar of planetary movements, people knew an oak tree should be felled when Mars was in the descendant.

The secret language of plants

Each plant is an enigma to be solved, a labyrinth of complexity we need to approach with care. Goethe drew our attention to this in his book on plants, speaking of a kind of primordial or archetypal plant whose template, as it were, could be rediscovered amongst the innumerable differences of form and behavior that plants develop in response to earthly conditions and diversity. Rediscovering this secret language leads us to a different understanding of medicinal plants and those used to make biodynamic preparations. Biodynamic agriculture is spreading fast in the realm of viticulture because – when properly applied, and when the plants cultivated by this means are fully understood – it has powerful effects on the quality and taste of wine.

Medicinal plants, including the vine, are ones which heal due to their very diverse, atypical characteristics. Sometimes such a plant will activate in its roots, stalk, flowers or fruit one or several processes 'normally' situated elsewhere. The beetroot, for instance, directs the abundant sugars synthesized by its leaves not into fruit but into its root! The maple directs these sugars into its sap or syrup. A medicinal plant heals, ultimately, by virtue of being atypical. And to understand why these medicinal plants heal we need to understand that human beings also contain these four states of matter. We can rediscover them in our four temperaments,[4] of which one is always predominant. No one will dispute that choleric people, through their stronger connection with heat, often seem to be on the brink of exploding (though they actually explode less frequently than one might think!). The sanguine person, always in movement, is as though borne on breezes and wafts of air. The phlegmatic benefits from his particular relationship to water and fluid through the inertia which enables him to navigate life's obstacles with good humor. Nothing suits him better than floating along contentedly in the swim of things! The melancholic rarely succeeds in overcoming the underlying note of sadness which his somewhat excessive connection with the earth, and the weight of matter, imposes on him. (Paradoxically, though, a good means of making him laugh is to tell him how sad life is!)

All these things are elaborated by Hippocrates, who says that an illness often begins due to a disharmony between these four states in a specific part of the body. In trying to observe – ah, such a difficult thing, and something a microscope could never manage, nor antibiotics ever render obsolete – where a process is out of balance, herbal medicine aims to offer a

---

[4] An ancient doctrine applied to medicine by Hippocrates (c. 460-c.370 BC). The temperaments – choleric, sanguine, melancholic, phlegmatic – refer to distinct personality types. See further in Gilbert Childs, *Understand Your Temperament!* (Sophia Books, 1995).

remedy by prescribing a plant or animal organ which innately embodies this specific connection or aspect. In this context a plant is no longer viewed as merely the bearer of substances or molecules, but as a gesture or potential link to the dynamic process it derives from its species. We can say that each plant offers a bridge between specific processes which our individual characteristics or conduct have unconsciously sundered. This is also how we can understand anthroposophic medicine,[5] which is still in its infancy, and, in a certain way, biodynamics too.

Observing the nettle

Let us take the example of the nettle to illustrate this in very brief outline, and to bring us a little closer to this whole approach. Though it may seem far-fetched, the nettle can help us, firstly, to better understand wine and its medicinal properties; and secondly to understand the vine and the choice of plants which help it accomplish its task in the face of an increasingly disrupted climate.

A quick glance at the nettle can show us that, like a sensitive and experienced diplomat, it seems to focus its activities more at its center than its extremities: its roots remain close to the surface and decline to go very deep into the soil; its flowers

---

[5] A holistic and salutogenetic approach to health that focuses on ensuring that the conditions for health are present in a person; combating illness is often necessary but is insufficient alone. The approach was founded in the 1920s by Rudolf Steiner (1861–1925).

which are spread the length of its stalk, rather than just emerging from the top, are both discrete and rather unattractive. As for its fruit it remains a minuscule, poorly developed, green grain. (see Plate 3.)

The Nettle

The nettle shows us, therefore, that it concentrates on its leaves: their transparent shimmer, almost floral in nature, and their burning sting, are qualities which ought not to be there in the normal run of things. We know that the leaf is the plant's mediating organ between its roots at one extremity and its fruits at the other. So it is easy to understand that the nettle is a plant which directs towards its center what many other plants express at their extremities. It is this fact of resembling something centered in its middle region, or, if you like, of balancing two extremes, which makes the nettle so beneficial for the human heart. This too is a central organ which continually tries to reconcile or balance what we load upon ourselves in our everyday lives, and mediate the often quite chaotic influx from our metabolic system on the one hand and our nervous system on the other. Rudolf Steiner[6] said of the nettle that it is an almost irreplaceable and indispensable plant. The vine receives it as infusion with particular gratitude at times of dryness or drought.

---

[6] Austrian-born philosopher and scientist who developed 'anthroposophy,' a system of knowledge that offers holistic and innovative approaches to education, agriculture, medicine, economics, the arts, and many other fields.

It can also be used as a liquid manure, just to improve soils, as long as its smell is not too strong: the soil dislikes a stench.

This is a very brief glimpse of the immense knowledge of plants which people once possessed, which Goethe first, and then Rudolf Steiner, founder of biodynamic agriculture, took up again so ably and developed further. Ultimately we are interested not only in plants' physical aspects but in their whole mode of life and the energies and gestures that inform them. This botanical body of knowledge has now culminated in the wonderful books by Pelikan[7] and Grohman[8] which give us very tangible and living insight into the profound and healing reality of plants without referring merely to their molecules. One can see each plant as a melody, and their totality as an orchestral ensemble and continually changing consonance. It is this 'globality' which can heal more than each individual note of the harmony, of which our modern, scientific world is often all too ignorant. The younger generation, hopefully, will engage more fully with these new ways of perceiving, which can sustain us and bring to life our dry and abstract view of nature.

The vine and modern schools of agriculture

We have now learned a little more fluency in the language which allows us insight into the profound nature of our friend the vine. Of course, like every plant it is pulled in two

---

[7] Wilhelm Pelikan, *Healing Plants* (Mercury Press, 1997)
[8] Herbert Grohmann, *The Plant* (SteinerBooks, 1996)

directions: by gravity towards the center of the earth, and by warmth and light towards the sun. The Greeks formulated this by saying that the vine is Dionysian in contrast to more Apollonian plants, and we can now grasp what they meant. An Apollonian type of plant is one which loves nothing better than to climb heavenwards with an uprightness that nothing can deflect, as if to rejoin the sun. Cypress trees, so fine and delicate, which climb to 10 or 20 meters, give us a perfect image of the Apollonian plant (see Plate 4).

The Christian religion chose wheat as the Apollonian plant par excellence. No one can really explain how a stalk as slender as that of wheat can climb so high and defy the wind. This is, indeed, only possible where a healthy agriculture allows silica to play its full role. Let me mention in passing that our modern types of wheat have, alas, degenerated due to the range of treatment they receive – including the dreadful height-limiting herbicides which stunt their growth to 60 centimeters. In former times they would grow one and a half meters high! They are then no longer able to 'disengage' themselves enough from the earth to connect with their destined energies, and cannot provide the human being with all the qualities which this plant originally embodied. It is now even accused of causing allergies, whereas in fact we ourselves have created them!

The opposite of the Apollonian type of plant is the Dionysian, which shows us a powerful predilection for earthly forces of gravity. Such a plant gathers in its roots a great strength for penetrating the hardest and poorest soils, and for making itself at home there! A stalk of wheat could never manage this. And yes, you have no doubt guessed it, the archetypal representative of the Dionysian plant is the vine (see Plates 5 and 7). In fact one cannot fully understand a vine without understanding its opposite. The forces which cypress or wheat direct into their branches or stems to defy gravity and elevate themselves ever higher and higher towards the sun, are directed in the vine to its roots, enabling it to delve ever deeper and penetrate even the most obdurate and stony soil. This strength also enables it to be quite thrifty in its need for nutrients. One can find vine roots sometimes as far as 30 or 40 meters away from the plant, or even more. Even if the soil is largely rocky the vine will take advantage of the slightest crack to insert its roots. This extreme connection with earthly forces means that the vine is almost incapable of accomplishing any heavenward ascent unaided. I do not say, however, that the vine resembles a liana – something often stated, but really quite inappropriate since the vine is much more than just this. As soon as its branches lift a little way above the soil, unless they have something to attach themselves to, they are recaptured by the forces of the earth. They therefore have to be propped and trellised, helped by means of wires or posts or dead trees – as people do in Portugal sometimes – to find the support which will allow them to climb upwards: something of which they actually have great need. When, towards the end of spring, one observes the almost desperate motions which the vine makes to hurl itself into the air, one could aptly call this a nostalgia for the sun. Each branch which falls is on the search for any aid, however slight, which will allow it to make yet another attempt to climb upwards again.

One has to understand that the vine is the earth's prisoner, imprisoned by the gravity which holds sway over it. This is described in the Greek myth in which Persephone, the daughter of Demeter, is kidnapped by Pluto – the subterranean god of the underworld who symbolizes gravity itself (and will-power). It is thus that Persephone becomes a prisoner of the harsh laws of the earth, like her son Dionysus who is subsequently torn to pieces by the Titans. They too symbolize the forces of the earth. Gravity is the force which imprisons what surfaces in the earthly domain, clothing itself in matter and isolating and thus separating itself from an overall context of energies. It individualizes in the physical realm, giving birth to the human being's sense of a separate self, of 'I,' that little word which, for better or worse, distinguishes us from each other. And this lost globality which we have to try to recover as human beings is, you can say, the second or reborn Dionysus whose heart is redeemed and entrusted to Zeus.

The vine is thus the archetypal plant of the earth, which joins in deepest union with it, accepting all its forces of gravity. Just look at its flowers, almost concealed in its bosom and often turned earthwards. Plants generally flower at the top, above the leaves. The vine is too drawn towards the earth to do this. To find the flowers almost hidden at its heart one has to move back its stems and leaves. But despite their smallness and discreteness we should not misjudge them: they generate a perfume one can detect from several meters away, thus showing that they retain a strong connection to the solar realm even if the earth condition dominates. Being a prisoner of the earth does not mean we lose all links to the solar realm of generating forces. On the contrary, it is this isolation which gives rise to a *greater longing for the solar realm,* as though by reaction. *The vine draws its capacity to create a product as noble and complex as wine from the fact that it is so radically atypical.*

And this allows us to raise an important question never tackled by our schools of viticulture, which are generally too far removed, unfortunately, from such qualitative considerations. As winegrowers is it our task, perhaps, to help the vine escape the earth a little, that is to give it some means to emerge from its terrestrial 'prison'? Do we need to help raise it up and become somewhat more Apollonian, with fencing or pruning that enables it to climb; or perhaps just to do this more subtly by spraying a Cypress infusion on its leaves? Or should we rather accept or reinforce this connection with gravity by, for example, cutting it back down each year when pruning, in order to force it to remain close to the ground? This, in other words, would mean stimulating and reinforcing its predominating temperament which nature gave it to enable it to survive in such difficult conditions. Couldn't such an action give the wine still more verve or vitality? A comparable approach can be found in education: to what extent should we indulge or oppose a child to help him achieve the best intrinsic harmony? How far can one go? In the same way the viticulturist needs to consider the age of his vines, the generosity of the soil, the latitude of the location, the direction of the slopes, the prevailing winds, hydrometric considerations etc., before taking a decision which will imbue his grape harvest with abundance, resilience or atypicity – unless he resorts to deceitful technology! Weighing up all these aspects each viticulturist will choose his own path, drawing on his *own creative response,* so as to better harmonize his vines with their archetypal forces. Let us simply add that each species of plant is naturally endowed with the forces it needs to complete its task. The key question is this: To what extent has the human being, in his agricultural practice, weakened this precious capital in recent decades through his lack of knowledge?

Shouldn't such things form an intrinsic part of agricultural courses? Should we not give students creativity and liberty by offering a much broader and profounder kind of knowledge than they will gain from fixed formulae or a mechanistic approach that is incapable of grasping the real nature of plants, and which instead focuses almost exclusively on economics and the market? In a location that is well-suited for wine-growing a wise agriculture allows the vines to fully unfold and express the profound life within them. Where a viticulturist can put his personal stamp on his vines through carefully chosen methods of cultivation, such creativity will certainly find its way into his wine. We will return to this theme in chapter 5.

Only an agriculture that takes full account of the laws of nature and its underlying forces, so widely ignored nowadays, can help generate the authentic diversity of expression implicit in each AOC. To think that this work should be undertaken at the cellar is, as we will see, a lack of understanding generally subscribed to by winegrowers who have had to transform their cellars into factories in order to try – imperfectly – to correct the agricultural errors of which they are hardly aware.

In general, giving the vine too easy and comfortable a time – that is, through too much manure, or excessively clean soils without competition from other plants, or planting too widely per hectare – i.e. always giving the soil too much strength – will nurture the vine's leaves excessively, giving rise to a rather feeble and spiritless wine. The vine will not have needed to resort to its full, powerful temperament, to its archetypal force, and so its harvest will not have received the stamp and imprint of its true nature.

These reflections, perhaps almost a little too detailed for amateurs, aim merely to explain why increasing numbers of viticulturists speak more of their fields than of their cellars, as if to underline the grave errors of a past which still very much haunt winegrowing today. Here we tap in to some profound questions: will our agriculture and our knowledge at last begin to ask of the plant – but not impose – a certain degree of effort? It is this human/plant synergy, and not an abstract and solely material body of knowledge, that enables us to participate in the genesis of a great wine.

I beg agricultural colleges to become aware of the arid intellectual state which they too often invoke in their students, the great degree to which they limit these students' creativity, the extent to which they distance them from themselves and their own human qualities. When, one wonders, will they start to teach students to observe life itself, or, for example, the way a vine leaf behaves, pointing earthwards from the moment it appears – quite opposite to, say, a laurel leaf which always points upwards (see Plate 5). We need to grasp that this gesture reveals the leaf's own profound nature.

Surely the time has come to teach students about the personality of each plant, thus enabling them to fully express themselves in a healthy agriculture and to produce foods full of living forces to nourish humanity.

For lovers of wine the important thing to understand is, first of all, the Dionysian nature of the vine, and secondly the need to respect this so that it can connect as well as possible with subtleties of soil and climate – so that, in other words, it can best marry its innate authenticity with the quality of the place where it grows.

A passionate viticulturist will be seeking for something hidden, for knowledge of the underlying forces which invigorate the vine and enable it to imbue its grape with as much force and elegance as possible.

Now that we have staked out the ground we must turn next to a subject fraught with trouble, and ask: What have we done to the vine over the past few decades?

2

Errors in Agriculture

Until the end of the 1950s not all wines were good, far from it, but almost all of them were authentic. Today one could almost feel nostalgia for the bad, real wine! Instead we continually encounter wines where the vine has been prevented from fulfilling its work, and where aesthetic intervention constantly has to labor to produce a good, but soulless wine by means of technology's thousand artificial measures. Here we find good, false wines, for which the AOC is really just writing on the label rather than a reality in the bottle. The full, original taste, which each 'appellation controlée' once guaranteed to the consumer, no longer exists!

How has this come about? What has happened? It is interesting to see how all this has been astutely orchestrated, and the extents to which producers themselves have been trapped in the process. The viticulturist was first approached with an apparently very attractive product, the herbicide. This was the *vine's first great drama*. A major and laborious task for winegrowers each spring and summer always consisted in keeping their soil free of weeds. (But nowadays the rules have changed as we will see.) This is tiring manual work undertaken during times when nature is flourishing without pause, so naturally a herbicide was tempting. Many tried it out. Those who advised it so insistently took care never to speak of the phenomenon, which scientists call mycorrhiza, so important for wine quality! What is this? Something very simple: to draw in nourishment a root needs the help of micro-organisms in the soil, rather as we need our hands in order to eat. *Each type* of micro-organism allows the root to assimilate *one particular geological aspect* of the soil. In their absence the root starves just as we would if we were sitting at a table full of food with our hands tied behind our back.

The commercial wiliness of the herbicide market consists in the fact that in just a few years it kills almost all the soil's micro-organisms. In the first few years this effect generates confidence in winegrowers since the death of these micro-flora fatten the vine. But after 5 or 8 years little life remains in the soil. If you dig a hole in a soil treated with herbicide for 15 years it is alarming to see how the life which goes to make up the soil has often almost completely disappeared. And of course the roots cannot live there either, and climb back up to the soil's surface (see Plate 6).

Chemical fertilizers

Let us translate the consequences of this secret assassination of the soil. When the vine can no longer gain the sustenance it needs from the soil, it becomes possible to sell viticulturists artificial growth agents in massive quantities. This is the *second great drama* of our AOCs, and one that is immensely lucrative for certain interests. A whole new market opens up, one already in existence previously of course, but hitherto fairly modest – that of chemical fertilizers – first for piling on the soil, and then for spraying on the leaves (foliar feed). There's no need to go into the details here, it's enough to understand that a chemical fertilizer is, first and foremost, a salt, and increases the plant's need for water. Just swallow a spoonful of salt and you will feel thirsty, and will need to drink to compensate for this excessive salinity. It is the same for all plants – for instance, a cabbage, once cooked, will give you back all the water it has absorbed.

And of course the effects of these fertilizers are not neutral. Apart from the fact that vines throughout the world, in all possible different locations, are fed in the same way, forcing a plant to absorb too much water is an invitation to nature – which always tries to restore equilibrium – to redress the balance indirectly through diseases. Microbes, viruses or fungal diseases should be regarded as heralds of imbalance or weakness. One day, I hope, this will be written in letters of gold above the gates of all agricultural colleges, which will no longer act as mere agents for the pesticide industry. Microbes and viruses should not be blamed for diseases in agriculture, as one would have us believe. They are merely the executors, the cleaners, like scavenging crabs in the ocean in charge of destroying all that is not sufficiently alive! In combating them without having corrected the imbalance that invokes them, one is just further weakening the whole natural system. To bring about a merely temporary cure, extremely dangerous molecules are applied, which isolate the plant *still more* from the living context it needs to attain its full health. It becomes clear that, in just a few years, we have created the conditions for a still more complex – and more lucrative – condition of incurable disease to manifest.

The 'systemic' drama

This slight digression is just to show that chemical fertilizers have led to a strong increase in fungal diseases such as oidium and mildew. These diseases already existed before, of course, but were infinitely less pervasive. The excess water that chemical salts force plants to absorb each time it rains, without respect for the growth that vines can draw from a spring season alone, calls forth fungi that try to regulate this excess water by taking up residence on the leaves. The first consequence of this is that treatments used in the past which were scarcely toxic when used in reasonable doses (Bordeaux mixture and sulphur) are no longer effective enough even when given in far greater concentrations which can, indeed, become toxic. Thus the door is opened to a *third drama* that is very injurious to the quality of our wines.

This is the 'systemic' herbicide. Having found and synthesized new molecules that are as dangerous as they are effective for suppressing the symptoms of these diseases which people doggedly refuse to understand, a process has been invented which *forces the vine to absorb these molecules through its sap.* It's a technical stroke of genius, for disease simply cannot appear! But in qualitative terms it's very serious. Let us try to understand the significance of what has happened.

Until the 70s these dangerous artificial molecules remained on the surfaces of leaves and fruit: rain washed them off, and the consumer could get rid of them easily enough by rinsing. This polluted rivers but not the food itself too much. The new process now forces these dangerous molecules into the very sap itself in less than an hour. Very practical, but these poisons – which are so toxic that the viticulturist who sprays them is legally required to wear a sealed suit with breathing mask – pass right into the plant's interior. It takes only a little imagination and understanding to see that, in the sap, they are going to pass into and help form our fruits and vegetables! The poisoned vine is supposed to eliminate these poisons within two months, if all goes well. But the toxicity of some of these products, used at a rate of several liters per hectare, is so great that just 2 or 3 ml placed in an egg would instantly kill a beech marten that eats it.

In addition, no one thought to tell winegrowers that sap is the vine's principal connection with the sun or the solar system, thus to all that gives rise to taste, life, the forces of ageing and maturation! It is evident that poisoned sap – and this is not putting it too strongly – cannot accomplish the same qualitative work.

Due to all these dramas, sometimes still heralded as progress, we have now arrived at a situation of abundant harvests, certainly, but bearing little imprint either of the soil that has been so weakened, or the unique climatic conditions which the sap does not properly absorb; and thus bearing little of its AOC distinction!

Cloning: weakening the natural expression of a species

This brief summary does not allow us to enter too far into further important details. But what we should understand also is the qualitative impoverishment that plant cloning leads to. Each type of vine – Chenin, Chardonnay, Cabernet etc. – bears thousands of original qualities unequally distributed in each wine stock. This is why, when replanting vines in the past, so-called 'massal' selection was used, which involved taking cuttings from hundreds of stocks, each one the bearer of different virtues, and using them to recreate a bigger population. In the case of each clone, by contrast, one sole stock was isolated instead, usually a very productive one, and the bearer of one or two clearly apparent aromatic qualities – but this overlooked the fact that beauty really only arises where a multiplicity of different characteristics create an equilibrium together, the one offsetting or complementing the action of another. This one stock was thus multiplied into hundreds of thousands or even millions of plants. Even if several clones per grape variety are now selected, the problem remains more or less the same. In addition, considerably increased yields have obliged growers to make 'green harvests,' that is, to get rid of a surfeit of grapes before they are fully ripe. These current practices, far from highlighting serious errors of selection, are nowadays often regarded as a positive development! In industrial-scale production, clones are seen as great progress. Deriving from the same originating stock, all these descendants flower at the same time and can thus also be harvested at the same time, even by machine. But from another point of view, that of full complexity of expression of the variety which each type of wine represents, this is real impoverishment. It is rather like a discussion on some subject or other in which either just one, or 30 people are involved, all with their different characters and perspectives. The monolog will certainly not lead to the same depth and scope. The clone which has reigned supreme for 30 years has impoverished the capacity of our grape varieties to express their full depth or abundance of taste.

Let us end this chapter by stating, quite simply, that a wine bearing little of the characteristics of its place of origin is hard to market if it has not been, let us say, 'overlaid.' Thus the gate has been opened to the *fourth and last drama of the vine,* whereby winegrowers, who had by now become prisoners of a system which often costs them between 1,000 and 1,500 euros per hectare per year[9] in artificial measures, were offered almost irresistible 'remedies.' Technology has thus come to

---

[9] Between approximately £650 and £1,000.

dominate our AOC wines, at the same time leading to their devaluation and destruction.

France, so rich in regional qualities and in micro-climate was thus poised to lose its loveliest adornment in exchange for selling its wines throughout the world and submitting to all the implications of global competition.

3

The cellar – the particular energies of a location; understanding the effect of forms

So much has been written on the subject of the cellar or *cave,* that in what follows I am going to emphasize what has generally not been said.

Let us first simply recall the role of the vinification process. The basic rule is simple: when the growth of the grape between spring and autumn is the pure product of nature, in other words when the vine, with all its originality of character and variety can, in its own way, transform meteorological influences into cellulose, starch and sugar without being interrupted by toxic artificial molecules, or artificial chemical growth enhancers, the 'must' possesses a kind of harmony or equilibrium which enables it to mature well in the cellar. All viticulturists who practice healthy cultivation can confirm this. Naturally there are still some basic simple actions to be taken, such as stirring the lees, and racking or decanting. But fundamentally, if you leave the vine to get on with its work, and still better if you actively help it in this task (and this requires a macrocosmic understanding of life as we will see later), if your vineyards are well situated and your grape varieties are well adapted to it, the cellar is just a kind of midwifery, nothing more nor less, to help things take their proper course.

There are, it is true, a few little secrets, and hardly more than that, relating to pressing times, cuvaison (maceration of the grape skins), or the period of contact with the lees. It is true that these actions have tangible effects, but overall there is little that needs doing in the cellar, except to observe with admiration what is underway. Nor is there much surprising or innovative in this. It is just a return to the situation which prevailed for many centuries, up to the end of the 1940s or '50s, when an agriculture that was still based on a balanced understanding of nature enabled the great vineyards to produce wines that still retained their majesty a hundred years later.

If, in contrast, one tries to replace nature without understanding her, if one thinks, as was still being taught a few years ago in the biggest wine-growing regions, that the soil is a dead substrate, of small importance, the wine deriving from a truly 'denatured' vine will become so precarious at the cellar that continuous, costly supervision and intervention will be essential. And there is really little point in astonishment at the unreasonable behavior of the must in the cellar, unless one also expresses astonishment at the often murderous agricultural methods applied to the vines themselves. It is only by accepting this latter view that one can really maintain that the cellar's work is essential to produce some kind of half-decent wine …

In fact one can characterize, and perhaps caricature, two different paths in viticulture: one which relies on replacing nature, and thus the AOC, obliging the cellar to fabricate a good taste, but without much characteristic distinction, and in which only one cloned variety can still often be recognized; and the other, in which, by hearkening fully to the vine and the life of the *terroir* where it grows, one allows it to unfold and flourish in the loveliest way possible. This just means that it can seize hold of the original qualities generated by a landscape, by particular animals, by various agricultural measures – all of which, as we will see, are rounded into a whole by biodynamics. Thus, as viticulturists, we can choose whether to become industrialized 'wine makers' or, instead, 'nature's assistants.'

The importance of climatic variations

To sum up our view of the right approach to adopt in the cellar, we therefore need to take a step backwards and examine in detail each one of our agricultural measures. We have already discussed how, from a non-material, intangible realm, photosynthesis creates noble matter in the grape. To deepen this aspect it is essential to understand how one can limit the work of the cellar. The climate is composed of three constituents which every year produce multiple variations. First the rain, either coming early, or slow to arrive, either abundant or sparse; then the heat, also subject to the same qualifying differences; and finally the wind, or air in movement, itself linked to the quality of light into which play the landscape's attributes: in particular, light-reflecting lakes or rivers. And all this diversifies any uniform continuity. Heat, for example, can quickly transform humidity into a passing mist or allow it to ferment in soils for a longer period. Winds participate in this task in a different way on each occasion. The Greeks had seven gods of the wind, of whom Zephyr was the gentlest. Yet again we see how past wisdom sharpens our faculties for perceiving subtle qualities and forces which in ancient times were embodied in the form of divinities. In a somewhat similar way the river Rhône in France is referred to locally as 'Doctor Mistral'!

And it is this rich diversity and complexity which affects soils, the mycorhiza and photosynthesis. Everything the vine creates each year as shoots, leaves and grapes is marked by these in-numerable subtleties that are never identical from one vintage year to another. And all this passes through the grape pulp in which the grape pips are enfolded, and contracts and focuses in them as in a synthesis. All plants work primarily to ensure the survival of their species, of their offspring. This genesis of the seed is an extraordinary thing. Goethe has revealed it to us in his poetry. After immersing itself in the physical world, and each year recreating itself wholly or in part, the plant has this capacity to contract to the minutest degree and to disappear into its seed – which will now bear its essence, all its knowledge and experience, until it unfolds again the following spring.

This process of contraction cannot properly begin until the plant has finished flowering. For the vine, so closely connected with the rhythms imposed by the four seasons, it is ideal if this genesis of the seed does not occur until after the midsummer solstice, when days start to grow shorter again and forces become centripetal.

The vine's self-expression

Flowering too early sets the vine at odds with the temporal rhythms with which it is so connected. Until the summer solstice, as days grow longer, the prevailing forces are centrifugal. In other words, as days grow longer and the sun's attraction stronger, nature increasingly expands, drawing the plant world outwards and upwards. These forces help the vine to construct its physical body. Only when the days start to shorten, and prevailing forces start to descend back into the earth instead, becoming centripetal, can the vine begin to contract and produce its fruit and seeds. Flowering too early runs counter to these forces. The vine has to start contracting to develop the grape and its pips, whereas the spring season continues to 'pull' the plant outwards.' This 'arrhythmia' often generates a wine that is promising at the first taste but very disappointing at the second. The fruit and seed will form, certainly, but will not properly bear within them the rhythms of the seasons. It would be interesting to organize a wine-tasting session on this theme – on condition, of course, that a healthy agriculture, unburdened by any technology, is allowed to participate. Thus the pulp of the grape can be seen as the 'bath' in which this seed forms, each time bearing in itself subtly different nuances. Steiner tells us that in the vine, some of the forces destined for the pip remain in the pulp, and that this explains its predilection for often explosive fermentation, and the poor capacity of the pips to produce a new vine stock.

All this goes to show that each year the vine is imprinted with the caprices or chaos of the weather, a little like a painter who is given different colors each time and produces a different quality of portrait upon his canvas.

The vine's labor is to use all the aspects of climate in a season, and all the variations of the life of the soil, to compose an overall, unified image. As artist of the earth it will always try to create coherence and harmony, even if certain elements imposed on it don't necessarily suit its preference. But to manage this, its archetypal force should not be hampered or obstructed by insensitive interventions. This is what goes to make a vintage. And this is why, even if the year and the climate have been poor, a vintage year can nevertheless be great. This is also why, for example, acidity or alcohol levels, which analysis shows to be excessive or inadequate, may be unnoticeable in the taste. This is also what allows wines such as Chenin to achieve a very pronounced maturity – where botrytis prevails – whose concentrating dynamic reinforces minerality. All these specific qualities will be included in the 'globality' of the vine's essential nature. Just as a certain somewhat vivid color can find its fitting place in the overall context of a painting (as the impressionists often show) so the totality, as it were, of vintages can come to full expression, on one condition: respect for life itself! This insight shows why official wine tasting sessions, to confirm that the AOC label is a true indicator of the bottle's contents, are actually so absurd. The requirement nowadays for obtaining the AOC designation is no longer the full expression of the AOC, but rather a 'correct' and perfectly nice wine – although one that has no location-related quality! With certain vintages the wine-growers most insistent about atypicity or technology impose their lack of understanding on their circle and surroundings in a way which one could humorously call a 'cosmetic AOC.' Those who try to return to the true AOC taste are often eliminated in the process. And this is why you will sometimes find in insightful cellars some wonderful wines that have to be sold as 'vin de table.' The courage of those who persist should be congratulated.

Fortunately this system of obligatory tasting, with all its limitations, is due to be revised and modified in 2008.

To sum up, we have to know how to help the vine to express itself, and to do so we first need to understand the extent of its capacities, so that we can keep our work in the cellar to a minimum. The first step in this direction is to practice organic growing methods. As we will see, biodynamics takes us still further.

The work in the cellar

The vital thing that I am trying to convey here, dear wine lovers, is that if one has not interfered in the work of composition that the vine itself knows how to complete during spring and summer – if leaves, flowers and roots have, through the system of energies that gives life to the earth, been allowed to communicate freely with their surroundings so as to respond to increasingly varied situations – and if, of course, yields have remained reasonable: if, if and if ... then each vintage will be worthy of the name without any technological aid to adulterate it!

But just imagine a situation, a very common one in fact, where the soil cannot nourish the roots, the leaves receive foliar feeds, where chemical salts force the vine to drink when it should no longer be growing, where poisonous molecules fill the sap which has to try to eliminate them as swiftly as possible, where 20 percent of leaves are removed artificially six weeks before harvest to 'aid ripening' (!), where the fruits tend to rot on the branch without ripening through lack of life forces and are sprayed three or four times with anti-rot treatment! Well, we will all agree that the 'work' will not be able to proceed with much coherence, and the behavior of the juice at the cellar will no longer be embedded in the wise, structuring forces which nature dispenses *freely* if we respect her. Thus the cellar becomes a factory and the viticulturist a 'wine maker.'

In this situation we are going to have to intervene all the time. First one needs to add yeast because the natural yeasts that arise in the course of the year have been killed off. One will have to go out and buy yeast that has often been produced by processes closely related to gene technology. In certain cases a wily commercial sense will even allow one to avoid this appearing on the label, since this isn't required if the added gene belongs to the 'family' of those already present. It is quite easy to imagine the commercial pressures that are brought to bear to obtain this kind of 'family right'! There is such a widespread and prevalent lack of understanding in viticulture today that it is often regarded as proof of professionalism to say that yeast has been systematically added to wines to avoid natural yeasts, which, it is added in all seriousness, 'lead to a bad taste'! Few winegrowers realize that when one modifies a musical chord with an alien element it becomes a dissonance; and that similarly, adulterating a wine with absurd practices leads to disharmony. Instead people are very easily tempted by the innumerable (over 300) aromatic yeasts, which in-depth economic studies have shown will persuade the consumer – of whatever social class, at whatever price range and in every country – to make a purchase. A consultant will be able to advise such things as: 'The banana-taste yeast did very well in Japan for 5-euro wines, but now try the one which produces a blackcurrant taste.' Yes, indeed, we are talking about AOC wines here, and not just table wines – which would be less shocking. The full palette of tastes is gathered here and made available to viticulturists, and can even be generated via a computer program. But, as you will agree, press articles revealing these practices are extremely rare! You will also agree, I think, that it is high time to act if we are to reserve the authentic qualities of the AOC.

Yet if, through lack of understanding, one has embarked on this path which will inevitably lead to smaller businesses collapsing, one has to become highly interventionist and continually fuss around the patient, keeping it alive with additions of yeasts and enzymes, through osmosis, temperature control, etc. One has to be permanently on guard to limit the aggression of all these living agents (bacteria) etc. which sense that a wine of 'poor birth' is their rightful prey, from which they *have* to nourish themselves in order to play the destructive role which nature has so wisely endowed them with. When life's proper equilibrium is destroyed or rendered perilous, the resulting produce is fragile too, and the aid of cosmetic adulteration or technology becomes indispensable. In contrast, by reinforcing the descent and assimilation of life forces into and by the vine – and this is what biodynamics excels at – the grape filled with these forces can cope with almost any of the viticulturist's eccentricities. You will now understand why people hold such differing views about the work which is needed at the cellar.

Fermentation

Let me give another example: what one could call the indispensable heat phase of fermentation which, for a few days, needs to climb to 26° or even 30°. This phase can be considered on the one hand either as potential decay, or on the other as an indispensable factor in achieving wine's full expression. Fermentation is like a fever, the search for a new balance through excess

of heat. It is not so long ago th at people regarded fever as a healing process. Today we fear or disdain it and try simply to dispel
it, thus distancing medicine from the healing forces which often
accompany it.[10] Fermentation has suffered the same fate, and
nowadays people dislike the bother of hearkening each year to
the different instructions which the grape issues about its state.
They limit the risks without understanding that if agriculture
has provided good conditions for the vines, the heat phase will
enable grapes to reveal additional qualities in the wine.

Osmosis

Nor can one overlook the role of osmosis, since it is now used
so much. This means concentrating the must by removing water, in order to enhance the tastes that the vine has not been able
to bring to full expression. This lack of understanding extends
to saying that this is done to suppress the effect of the last rainfall, considered undesirable. But this is to overlook the fact that
when a soil is not full of chemical salts – if the earth filters its
rains through proper microbial life, and if the roots of the vine
have not been curbed in their descent by these dreadful synthetic molecules, and can delve downwards for meters – rainfall
will very rarely hamper the vintage. And even if 100 millim-
eters of rain has fallen in a few days, biodynamic methods can
still, certainly, achieve an interesting wine, obliging the vine to

---

[10] Of course, fever can be dangerous to a patient if it rises too high, but
a moderate fever can, as symptom of the body's own self-healing forces,
accompany and aid the healing process.

exhale this water by treating it with a quartz (silica) based remedy. If agriculture is healthy, a vine and its grapes will not gorge themselves on water but will drink *reasonable* amounts. It really isn't normal for an AOC wine to be subjected to osmosis; but if this is done (being legal nowadays), 'wine concentrate' ought to be printed on the label, as on cheap orange juice. What consumers can assume is an exceptional year, in which dry conditions and sunshine have married well, is sometimes just due to reverse osmosis which, let's not forget, reverses the wine's natural polarity – and with what consequences for the ageing of the wine? It also opens the door to excessive yields. The aim here is not to examine every single detail of this almost industrial-scale viticulture, which even sometimes affects very costly wines and which spreads a little further every day thanks to the adroit hand of certain key people at the Institut National des Appellations d'Origine (INAO),[11] who are more concerned with their own careers than the wine itself. What I am endeavoring to do, rather, is to explain in fairly simple terms to people of good will that all these artificial methods are chiefly linked to ever-increasing agricultural errors.

The micro-oxygenation method

Let us, finally, take the example of micro-oxygenation. This new hobbyhorse that has become useful to certain winemakers just replaces the work that the leaves of the wine

---

[11] National institute of the 'Appelations d'Origine.'

should do for the grape if agriculture was healthy. And in this latter case the practice is useless. All these increasing errors have their consequences in the cellar. The centrifugal forces will try to expel synthetic molecules that have been forced into the sap, and in doing so will destabilize the wine. The enzymes nourish the yeasts that are either weakened or not adapted to the specific conditions of the year, etc. etc. I do not wish to condemn a producer who chooses this course, but only to tell him that as far as wine technology is concerned France really has nothing to export. Too many countries with readily available cheap labor can achieve exactly the same taste at a quarter of the cost or less. And if we fail to understand this, commercial collapse will follow swiftly. No region is protected from this danger! Let us not forget that in 2005 the government of France had to dispose of 1.5 million hectoliters of AOC wine that had not been sold. In addition, all these practices and many others tarnish the image other countries have of France and its wines a little more each day. Our diversities of climate and geology, perfectly adapted to our grape varieties, and linked to the perfect adaptation of our vines that have sometimes been growing in our regions for thousands of years, ought potentially to ensure that we are almost beyond the reach of global competition. I don't wish to imply that French wines are the only ones worth considering, but in France we do have the resources of a viticulture which is, rightly, the envy of many others.

A way forward

Fortunately another kind of viticulture exists, still very dispersed, but one which increasingly understands the traps that have been set for it. It is returning to 'real wines' where the word 'terroir'[12] which has become so debased today, is regaining a little of its true meaning. In fact, in allowing all these artificial measures, in using them too much, a space has, paradoxically, been opened for those who want to go in a different direction. The less credulous customer is starting to discover the truth that has been hidden from him. Is it really worth going into ecstasies about the 'nose' or the taste of violets in a wine if one knows this has been added through aromatic yeasts? This question is one which needs to be asked clearly! To accelerate this return to real wines it would be enough to require by law the publication on every label of the technological practices involved in producing the wine. For example, this might read: 'wine concentrated by osmosis and perfumed with artificial yeast X, which gives a nose of violet.' Of course this will never happen, since too many interests are at stake in the commercial wine world. But it has been possible, by contrast, to form a group which gives a *legal guarantee of the authenticity of the taste of the AOC*. This group has existed for five years now and today includes 140 viticulturists in 10 countries and has met with great success in all continents. Its charter of quality is the

---

[12] Translated literally as *soil*, this French term describes not only the terrain on which the vines are grown, but also encompasses soil, slope, orientation to the sun, elevation and effects of climate.

strictest in the world. You will find details of this association in the Appendix and at the website www.biodynamy.com.

In speaking of real wines we need to give a very precise definition of this, with a very strict charter of quality – also in the Appendix – which lists all the actions prohibited both in the vineyards and in the cellars. Such a charter will give consumers not just virtuous words, as is often the case nowadays, but *a legal commitment on the part of the viticulturist* relating to his vines, his wine and the consumer. Subsequently it is necessary for the wine to achieve a certain level of quality and originality. The group's aim is to show, 'glass in hand,' that in different locations grape varieties express themselves differently in each case, and that it is these subtleties, unmasked by cosmetic measures, which endow a wine with its own innate charm. A wine should not just be 'biodynamic' but it also needs to be good, imprinted with the originality of the place it comes from, and by the labor of a well-adapted grape variety.

On the path towards real wines there are of course very different levels of commitment and achievement, with fairly recent dates of conversion to organic or biodynamic agriculture, and widely varying results. Sometimes the will is there but not yet the results. It is common to hear those who want to get on board this new market without taking rigorous decisions saying that they are '95%' organic or biodynamic. Just a moment's reflection will show the vacuousness of such a phrase. As James Milton, a remarkable biodynamic viticulturist in

New Zealand, joked: 'Can you say a woman is 95% pregnant?' One also hears people saying that they use very few artificial herbicides, just enough 'to save vines from disease'… without understanding that it is these same herbicides which over time accentuate the spread of disease. People speak of such agriculture as real progress when in fact it still keeps using chemicals. This approach, certainly, represents a praiseworthy effort – and a cost – for the winegrower. But it does not allow the vine to develop properly. It is a little like only half-tuning an instrument – the music will not be harmonious. A winegrower willing to make an effort should simply first try out a hectare or half a hectare with organic or biodynamic methods (the latter has more rapid results) and see after 2 or 3 years what qualitative difference he obtains.

In all these cases, there is increasing awareness of the harm caused by synthetic molecules used in conventional agriculture – already progress in itself. Often all that is lacking is the knowledge necessary to avoid diseases intrinsically connected with health deficiencies. Nowadays no school of agriculture is fully able to teach such knowledge. Organic practices taught today consist largely of replacing a synthetic molecule with a natural one! Even here, what is retained is a mechanical understanding of life, rather than discovery of a vast system of intercommunicating synergy.

How can we swiftly repair the immense harm done by the last two decades of agricultural practice, whose dangerous effects on human health are becoming ever clearer to scientists?

It is no longer even possible to claim that this form of agriculture will help avoid famine; on the contrary, it creates it since many poor countries cannot afford the increasing cost of the chemical products required by this approach. It would indeed be possible to feed the world in a healthy and more economical way – something wholly removed from a party-political stance, since the laws of life do not belong to any political movement. In viticulture the question is how to rediscover benign treatments based on natural, healthy plants and products. I say healthy since one can also use natural, and therefore organic poisons, which is not advisable. How can we give back to wine, tea, coffee, chocolate and vegetables a taste imbued with the place where they grow? All such questions lead us to biodynamics.

A return to indispensable sources: the science of forms

Before going into further detail about this new approach, let us stay in the cellar for a while longer to speak of shapes and forms, which have wide-ranging implications. We know how barrels can affect the taste of a wine, but unfortunately we are mostly unaware of its principle quality of shape. The barrel is based on a 360° vault-shape, passed down to us from a long-distant past when people better understood what underlies life, and thus the different energy effects generated by each shape and each place, in accordance with proportions, matter etc. It is after all only very recently in history that humanity has begun to plunk houses down anywhere, with terribly arbitrary shapes. A house is also the place where we sleep. And when we sleep we are no longer protected by our individual self, by the 'ego' that temporarily leaves the body to allow it to recuperate. During sleep we are therefore fully subject to the force-field which the place and the house generate, for better or worse. We need to understand that when we are sometimes unable to sleep in a house this is, in fact, a protection given us by our individuality so as to avoid our exposure to forces which could be very harmful. Something similar is true of the vine which, as it grows, also needs to receive a whole spectrum of living forces. In olden times no one would have made much of a song and dance about such knowledge. Of what do these energies consist?

Gravity

A long time ago the Romans had a custom of grazing sheep for a year in a place where they wished to build a house. Then the sheep were killed and their liver was examined to decide whether building should go ahead. Let us try to understand the very specific reason for this practice.

On earth, as we have seen, everything is balanced between the laws of earth and its gravity which weighs us down, and the solar forces which uplift us.

One can put this in a more scientific manner: everywhere on the earth there are places we may call 'absorption' points, and those where gravitational forces are stronger. Gravimetric maps show this clearly. Gravity, or weight concentration varies everywhere on the earth, increasing as we pass from the equator to the more flattened poles. There is less gravity at high altitude, of course, whereas the density of subterranean conditions increases it. It is very strong in rock clefts or crevices. It varies each day depending on the positions of sun and moon and their distance to the earth. For instance, it is important to know that, due to all these facts, cited here very approximately, the sea is not a flat plane, and that the water in it creates depressions and swellings differing by over 100 meters – resulting from the different forces to which it is everywhere subject. Subterranean conditions, with their folds, compressions and geology, naturally play a part in these influences which we find everywhere on our planet. Thus all such things count in our understanding of the laws of living things, upon which depend the health of vegetables, and our own health too. This is an understanding we can acquire over time when we practice biodynamics. But gravity is not the only force active on earth. The earth is, in fact, covered with energy 'corridors' where different natural forces – i.e. electro- telluric, magnetic, even radioactive as in Prague – encounter each other and interact, rather like the veins in the human body.

If one is interested in the culinary arts, in diverse, subtle tastes that can be found in both food and wine, one almost has a moral duty to penetrate the secrets of this system, which in such diverse ways helps enhance aromas, tastes, colors and odors. Such an understanding will help us approach viticulture differently. It is true that this path is a good deal more laborious than just using modern technology's artifices to perfume a food product which has been unable to integrate and embody all the qualities for which it has the potential. What does this system of energies consist of, and how does it function?

Electro-tellurism

The term electro-tellurism is often poorly understood, and refers to the difference of voltage resulting from polarization, in other words from the different degrees of life's incarnation in or penetration of matter. No matter exists without electro-tellurism. The electric currents constituting it are very weak but everywhere present nevertheless, and indispensable for life on earth. No cell exists without electric charge. If you tie a metal thread from the top of a tree to its base, you will risk killing it quickly because you will destroy this difference of voltage. This electro-telluric system is active everywhere in the life of the planet, and varies constantly as it is influenced by the links between the earth and the solar system, and even by the continuous activity of the sun. All is constant flux in the energy realm. People have even thought of using these ubiquitous currents for communication during times of war. Alas, what we today call 'vagabond currents' are the result of charges escaping from high-tension cables, power stations, rail lines etc., in fact from all the activities of our daily life. This considerably pollutes or disturbs the natural organization of this electro-telluric system which is essential for life on earth, and few really give heed to the seriousness of such pollution for human health. We need to deepen our understanding of the system which organizes all natural life on earth, of which, by the way, many doctors are aware.

Magnetism

Magnetism is also an essential force indispensable for life on earth. It orientates the needle on our compasses, entering from the north and departing from the south of the earth. Under certain conditions magnetism can modify substances, magnetically polarizing them; we can then speak of the 'magnetic memory' of a place. Magnetism also forms part of force fields important for the balance of life on earth and varying everywhere, like electro-tellurism – to which it is connected – depending on the nature of local geology, the metals which the ground contains, its shapes etc. As in the case of gravity we can buy maps which show detailed magnetic fields present on the earth (similar to those showing seismic patterns, which are also important factors). For magnetism there are also zones of emission and absorption, and sometimes even of inversion of the north-south orientation – i.e. reversing the planet's global orientation. The earth's magnetism is closely linked to that of the sun which regularly undergoes magnetic inversions following cycles which are not always understood. One of these, lasting 26,000 years, is what the Hindus used to call the inhalation (centripetal forces) and exhalation (centrifugal forces) of Brahma. This corresponds wholly with what we know as the Platonic Year (25,920 years comprising 12 times 2,160 years) whose deeper meaning is explained by Steiner.[13] This is the time it takes for the sun to rise at the spring equinox before every single constellation of the zodiac, in other words to complete a rotation of 360 degrees.

Solar eruptions create magnetic storms on earth which, for example, aviators are very familiar with in the Bermuda Triangle. People suffering from cardiac disorders know the extent to which the heart is connected to magnetic storms. If one increases a person's magnetic field while his electrical field remains constant, the heart contracts. In the reverse case the heart dilates. Yves Rocard, father of the former French minister, was a doctor and mathematician, and wrote a reference book in which he studies the effects of magnetism on the human being. All these facts show the array of interconnections in the solar system to which we belong, and the extent to which microcosm and macrocosm are continuously connected. It appears that the center of our galaxy also plays an important role in relation to earth's magnetism. This whole electromagnetic system is rather like the earth's nervous system.

---

[13] See, for example, *Mystery of the Universe*, London 2001

No chemical reaction is possible without electromagnetism. All such reactions are accompanied by altering electrical charges. These are the electromagnetic fields which 'link' and 'unlink' the constituents of living bodies. DNA and its double helix of left and right (clockwise and counterclockwise) form a kind of antenna for these electromagnetic fields. Lakovsky stated this at the beginning of the 20th century, before it was confirmed by physicists. It is interesting to discover that the electromagnetic field of every human being has its distinct characteristics, forming a real, individualized weft of energy that is sometimes called the 'aura.' The so-called Kyrlian effect, described in several books, enables us to form an image of these forces. It is essential to understand that these electromagnetic fields allow living organisms to receive and express their life forces. They are, to some extent, the vectors or bearers of life on earth. In line with Lakovsky we can say that all organisms on earth resonate with forces from the solar system, and even from deeper in the universe, which bombard us ceaselessly. Living organisms however, are not just receivers but also *emitters*. This is also why the loss of a plant variety or species of animal is a real qualitative impoverishment for the earth, a little like a failing television set that can no longer receive all the channels. This is an impoverishment of the communication network which can be said to sustain the earth. The whole earth itself is also a receiver/emitter, as are the planets. All this has been proven scientifically; the only thing lacking now is to link and create a synthesis of all these discoveries.

Now it is hopefully easier to understand the fact that in certain places we can discover 'specific energy conditions' that are very powerful and unusual. It may initially seem surprising that precisely in such places we often find exceptional architectural works, or more ancient megaliths – which are equally powerful in terms of the energies they release. All this is nowadays studied as part of the discipline of 'geobiology' which links biology – that is, the forces of life – to a geographical location. This is sometimes also referred to as 'sacred geometry.' Some of these specific points – where unique works of art have often been created – act rather like the earth's acupuncture points. By constructing places of meditation or prayer there (for instance Chartres or Mont St. Michel) or places of academic or artistic study, or theaters such as those of antiquity, human beings can, in different ways specific to each location, engage with and draw on this complex system through their search for truth, their state of awareness or devotion, their songs and religious rites. In such an instance the human being sustains the earth as much as the earth sustains him.

To deepen our understanding of viticulture, of the work accomplished by the vine and the profound significance of AOCs, we need to rediscover the whole impact of this energy organization. An AOC with its original qualities and tastes is, in the end, nothing other than the visible culmination of all these factors which 'interlace' on the energy plane before finally giving rise on the material plane to subtleties of taste.

Before attaining an ultimate physical state there are millions, even billions of continuous permutations of atoms and cells that, in mutual sympathy or antipathy, arrange themselves in one way or another to form a plant, an animal or a human being. When embryology examines all these mutations occurring in a fetus to create a living body it is hard to overlook the workings of this invisible but very active system. This genesis also contains therapeutic secrets whereby every organ retains a kind of memory of the 'encounters' it underwent as it was developing. This is somewhat like the memories we retain of our neighbors or friends when we were children. Certain approaches to medicine – such as anthroposophic medicine – intentionally draw on these connections between an organ and its originating sphere of resonance. The laws governing all such genesis contain profound realities. If we fail to study the underlying planes that sustain life but instead hold fast to material things alone, this is the same as limiting the analysis of an artwork to what the microscope might tell us about it.

All Celtic art shows us how people used to sense the interactions of all these energies in often opposing polarities. We can find this in the extraordinary *Book of Kells* in Dublin, which cannot simply be regarded as a work of art. It really illustrates the forces which bear life into the physical world, reproducing these before our eyes in an interlacing dynamic of life, a weaving interplay of energies.

George Bain, Celtic Art, Glasgo 1951   Dublin, Trinity College

Book of Kells

A book by Rudolf Kutzli, *Creative Form Drawing*[14] enables us to go a good deal further in experiencing how forms change into each other. This also allows us to better understand the continuous interaction between centripetal and centrifugal forces – i.e. those descending into or expanding out of matter – from which all the forms surrounding us arise.

When people in olden times constructed cathedrals, fortified castles, monasteries and even sometimes wine cellars, they were very receptive to the interaction of certain of these currents of energy. They took advantage of all such effects, enhancing and 'ennobling' them.

---

[14] Hawthorn Press, UK, 3 vols

Ennobling terrestrial forces: the knowledge of builders

But how was this enhancement achieved? Ultimately this question affects all the sciences: how can one help a macrocosmic world of energies to enter more deeply into the material world so as to order and form it? How can we accentuate the presence of vital energies in a particular location? This is the key question that today's architects have forgotten, too often losing their way instead in mere intellectualism. These approaches and insights of the past contained a powerful intuitive knowledge which some scientists are now patiently reacquiring and quantifying.

Connecting to archetypal forces

It is essential to understand this more tangibly, at least in principle, to help us better draw on such forces and intensify them as biodynamics does. In essence, as I hope is clear by now, life is composed entirely of forces. How, in the past, did people draw on these very diverse reservoirs of energy? How did they reinforce their descent or presence in a particular location? This is actually a question very close to the one that biodynamic winegrowers ask themselves when they try to help the vine unfold the full expression of a region's specific qualities, that is, its soil and climate. It is this insight that will allow farmers and winegrowers to rediscover the nobility of their profession.

In architecture everything starts from the circle or from concentric circles, one generating another (as we can see when we throw a stone into water), and all resonating outwards from the same center and point of departure.

This brings us to a question often ignored, or approached with too much of the cold intellectualism that harms so many of our children: what is this central *point?* Euclid gave the following striking definition of the point: 'A point is that which has no part.' Yes, indeed, we need to grasp the grandeur of this answer and its implications. The point has as yet no surface. One could say that it has not yet conquered space: it is symbol or first inkling of a birth in the world of matter, the physical world which surrounds us. Thus the point is the necessary entrance through which the world of energies everywhere latent behind every material particle can descend into matter. It is the passageway through which forces must pass into the material realm, where other laws prevail. This can also be expressed as the place where the two circles that form the figure 8 (also known as the lemniscate) touch each other and where *their direction is reversed.*

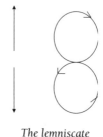

The lemniscate

This can lead us to a better understanding of what physicists describe when they speak of the void not being empty but full of something else. This is the hidden significance of the horizontal lemniscate which also, quite rightly, symbolizes infinity. It refers simply to the active world of highly organized energies which connect the earth to its living forces or dynamisms. The 'point' is *the precise moment* where they enter the physical plane. We need to grasp this very clearly, for it tells us nothing less than that the origin of living things is not of the earth itself, but approaches from the wider universe, from our solar and stellar systems without which all would be corpselike and dead. This knowledge of energies is something that our age has, whether through dogma, lack of insight or profit motives, banished from the realm of science. It is not possible to cure the cause of a disorder without first studying the system which recreates order. Any illness, whether of plants, animals or humans, derives solely from lack of organization! Simply by studying the system which creates coherence we can gain access to the root causes of an illness. Yet in France, in particular, too many excellent doctors ahead of their time have been removed from the list of registered physicians for having dared suggest that disease involves more than merely material factors. The profundity of the Hippocratic Oath and the double spiral of the caduceus – which is still the symbol of the medical profession – are no longer properly understood. We find the same symptoms in the domain of agriculture. A great agriculturist, Clause Bourguignon, has suffered the ire of the INRA association to which he belongs for daring to speak the truth in his book *Le sol, la terre, les champs* ['The soil, the earth, the fields']. Jean Pierre Berlan, the author of several books, has been marginalized for daring to denounce the potentially grave risks of genetic engineering – which overrides the archetypal regulating forces that organize terrestrial life.

What errors and astronomical costs could have been avoided were it not for the dead hand of economic imperatives upon the laws that engender life. Fortunately, little by little, we can see physicists, astrophysicists, doctors and researchers at CNRS (French national society for scientific research) developing a new knowledge based on insights into the weaving energies that sustain the earth. If this knowledge can be allowed to develop without constraints it may be able to alleviate the vast problems facing medicine, agriculture, education, and even society as a whole.

The Platonic forms

Let us return to Euclid for a moment. From the 'point' viewed in this way we can develop the continuous line which is simply a sequence of such points; and can then come to examine the forms, outlines and generative levels of space. Of the straight line Euclid tells us: 'A line is a length without breadth' – which may now be a little more comprehensible. Yes, the line is simply a sequence of points, a discreet trajectory of entrance into the material world since this line does not yet have breadth. This shows us how the sensibilities of Euclid's time were different from our own. How deep they were! From the straight line, subsequently, we can develop the radius, the circle and all that derives from it, and in particular the extraordinary polygonal Platonic forms consisting of 5 so-called Platonic Solids that are so overlooked today. They are composed of the tetrahedron (four planes) the hexahedron or cube (6 planes), the octahedron (8 planes), the dodecahedron (12 planes) and the icosahedron (20 planes).

Why are these forms important, even essential? *All living creatures of 3 dimensions result from combinations of these five, simple forms.* Historians tell us that this was known well before Plato, in an age where human sensibility had not yet been

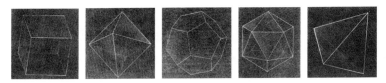

The Five Platonic Solids.[15]

replaced by our expensive scientific instruments. Thus, in the living world which surrounds us, we can invariably discover a geometrically organized genesis of forms giving rise to the triangle, the cube, the pentagon, the hexagon and their multiple interactions. The forces of earth crystallize or coagulate along these intersecting planes in material atoms that become visible

---

[15] In *Platonische und archimedische Körper, ihre Sternformen und polaren Gebilde*, Paul Adam/Arnold Wyss, Verlag freies Geistesleben.

to our senses. We can also say that this is where centripetal forces exert their influence. According to the 'rhythm' of connection between the exterior planes of these Platonic forms we obtain simple or indented forms. These Platonic forms are continuously active in the mutual interaction of atoms, in the division of cells, in the genesis of all life on earth. In the realm of the microscopically small we can see these pentagonal, hexagonal or cuboid structures everywhere.

These five solids are all connected to the circle since, extended to the infinite, they always become a circle or sphere. Thus we can say that they potentially rejoin the unity that incarnation into matter destroys, retaining a specific kind of connection to this globality. In their distinct and original geometric form each one is, in its own way, the bearer of particular precious forces, indispensable for life on earth. Their forces surround us always and their effects impinge on us, yet they remain invisible.

Each of these Platonic forms bears the archetypal forces of a particular planet: Saturn for the cube, Jupiter for the tetrahedron, Mars for the dodecahedron, Venus for the icosahedron and Mercury for the octahedron. See the bibliography for volumes that explain the full complexity of how their energies impact on earthly life, and also the transformations from one to the other that we see continually illustrated in embryology. We can say that these are the key forces of life on earth, as though geometrically 'summarized' or 'reunited.' All these forms call on and link with forces acting as a kind of antenna for them. The great masters such as Leonardo da Vinci and many others reveal the connection between human beings and these forces.

In these designs, ancient authors show us the human being's physical body to be the result of all planetary and stellar forces, with lines and key angles that touch every part of the body. Every angle, plane or proportion reveals a connection with different forces. The sum of all these diverse influences creates the extraordinary architecture of the human body.

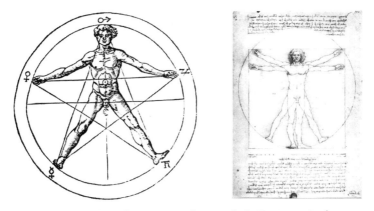

The human being by Agrippa and Leonardo da Vinci contained within the Platonic forms

A little reflection enables us to rediscovers all around us the effects of the forces of pentagons, hexagons, squares and triangles in the shapes of leaves or flowers, minerals etc.; and also in the interior of fruits (see diagram on the next page and plates 7 and 8).

This system surrounds us but we no longer perceive it. Matter is ultimately nothing other than energy concentrated by

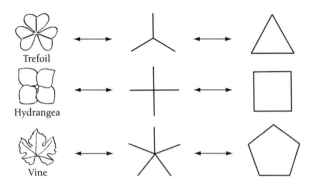

*Pentagonal, square and triangular crosses alongside
leaf and flower forms*

gravity and maintained in this state of contraction by electromagnetism. To put it more poetically we might say that matter is merely a winter coat, a necessary cloak growing visible where the laws of gravity are stronger than solar forces. And we try to persuade ourselves that genes – docile servants of these forces – are the origin of this life-giving system.

We can also rediscover the wisdom latent in the forms of certain musical instruments, for example the violin. Its harmonious proportions are as though 'tuned' to precise archetypal forces. Drawing on this fact it creates a perfect vessel that resonates with a world whose essence is composed solely of vibrations. Ultimately we can say that sound generates form. For instance, we can see how it gives shape to iron filings on a piece of paper. All older civilizations knew this. But in return, form also generates sound, as musical instruments ceaselessly show us. We could say, a little more scientifically, that an instrument whose form is born of harmonious proportions generates greater coherence in its emission of sound waves. Forms have the power to modify sound waves, as do metals: a trumpet made of copper or silver will, as musicians know, have different kinds of sonority.

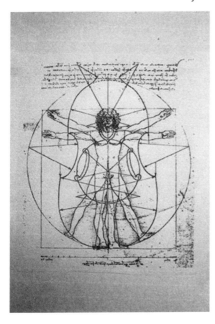

Sacred proportions in a human being and a violin

We can now extend these insights to many other shapes and forms that surround us. The shapes of furniture, for example – the harmonious design of certain canopy beds was meant to guarantee a perfect night's sleep. Here we can also rediscover the meaning and purpose of numerous protective seals, sometimes called talismans, whose diverse geometrical shapes created a connection with different specific energy forces. A science of energies underlies all this. In fact, the advertising world draws on such knowledge of forms and their secret resonances within us when it seeks to invoke in us a 'gesture' of purchase.

For viticulture and agriculture, and even for culture in general, what this means is that we can try to help people create a conscious connection with active forces of which we hear so little in our schools and colleges. This impoverishment has very serious consequences: people speak only of matter and never of specific latent energies working upon it. When studying a form, for instance, we must always try to see its complementary aspect – that is, what surrounds it – which is the matrix out of which the visible is born.

But at the same time we should be careful not to get stuck in the past and keep repeating it. Everything is evolving in the universe, and the forces active on earth are also changing. Nothing is fixed, and this insight allows us to change and develop. For example, pyramids – even though they are astonishing structures – are no longer appropriate to the life and consciousness of modern human beings.

*Gothic seals, which we can see are derived
from a deep knowledge of forms*

To conclude this simple introduction to the Platonic forms we should remember that the earth is situated at the heart of the solar system: between the exterior planets Mars, Jupiter, Saturn – which have a longer year than our own – and the interior planets (Venus and Mercury) whose year is shorter than ours. In the same way, to unfold their full activity, these Platonic forms also need to be placed one within the other like Russian dolls. This gives a potential source of energy and harmony with the power of profound healing for human beings.

Outermost circle: Saturn
    cube
2nd circle: Jupiter
    tetrahedron
3rd circle: Mars
    dodecahedron
4th circle: Earth
    icosahedron
5th circle: Venus
    octahedron
Innermost circle: Mercury

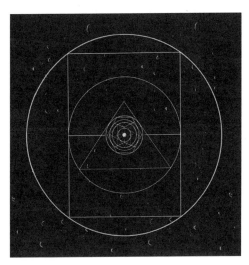

Platonic forms, their polarities and dominant planets

All great architectural works of the distant past, wherever they are on earth, bear the signs of this profound knowledge. The Goetheanum, built in wood by Rudolf Steiner at the beginning of the 20th century, then redesigned and rebuilt in concrete after it burned down, is also derived from the circle and pentagon: from opposing forces and a mediating force that joins and harmonizes them. We should not forget that the more strongly two forces oppose each other, the greater will be the health-giving effects generated by their harmonization. One important thing which Rudolf Steiner re-introduced to his architecture from the past was the notion of the metamorphosis of forms, or the passage from one form to another, at the same time creating a link between art and science. We can regard this in a certain sense as 'architectural embryology' or a living, organic architecture which calls forth inner participation from those who experience it. As we connect with the form into which we enter we can try to 'be the form' and thus inwardly experience its transformations as we pass through different spaces. This is a little like the transformations of a seed that passes through diverse forms before reaching its fully-grown state. Such architecture is an important evolutionary impetus.

It will hopefully be clearer by now that all the details of great buildings of the past correspond to very precise geometric rules where nothing is left to chance. In all such buildings the pillars represent a place of interaction between the two opposing forces of gravity and solar attraction. To put this in another way, this is where telluric and cosmic elements connect or change into each other. This knowledge was already present in ancient dolmens and menhirs. We can see it more clearly in Gothic art, for example in the choice of carvings at the summit of each capital as we progress towards the center of a church. These are not, as is too often imagined, merely symbolic decorations but a tangible illustration of the forces embodied there. Who still has the knowledge to attempt such things today? The architects of great, sacred buildings of the past knew how to sense, seize hold of, master, concentrate and give physical form to macrocosmic forces in a specific physical location and space. In those days the word 'religion' – derived from the Latin word meaning to 'reconnect' – implied that human beings could pass into a place which generated a particular context of energy forces, allowing each person to feel less isolated, and more connected to the universe. Such places – Buddhist or Hindu temples, Egyptian pyramids, cathedrals such as Chartres in France and Naumburg in Germany, or inspired Celtic churches in Brittany, all possess their own distinct relationship with energy forces and generate qualitatively different atmospheres, as everyone can sense for themselves. This understanding, with its health-giving potential, is something sadly lacking in our modern world of arbitrary and functional spaces.

Some forms comprise precise proportions of breadth and height that generate very beneficial effects for the human being and life. An arch corresponding to these laws always re-establishes balance in relation to the ground's tug of gravity. Houses under which the negative energy of a stream flows used to be constructed with an underground vault to allow life to remain healthy above ground. An underground stream generates negative energies because it disturbs the balance between negative and positive ions and creates radon gas. This radon phenomenon can even lead to death where the direction of the stream is opposite to that of the sun, i.e. West to East. No cathedral was ever built without several water courses in the foundations which all cross under the place where the altar is situated, and above which an arch is constructed. The rising negative current is reflected at ground level by the vault in such a way that these two negative aspects become a positive one. One could also mention the labyrinths designed by Leonardo da Vinci, or Dürer, or, again, Chartres Cathedral, which highlight this mastery which once existed over the world of energies always present behind the material world (see Plate 9).If you talk to estate agents they will tell you that out of 50 or 60 houses there are often one or two which are regularly put back on the market. They are located in places with excessively negative energy. Some people, particularly women, have retained this sensitivity which allows them to feel either better or worse in a house, depending on its location, its forms and materials, and even its past history. All this belongs to the realm of 'geobiology,' and is a field of enquiry essential for those who wish to gain a deeper understanding of biodynamics.

The law of harmonies

These considerations lead us towards a rediscovery of precious laws of resonance and harmony, acting as the bearers of specific, regenerative energies. It is by means of these laws that we can find a degree of release from subjugation to our separate individuality. Just think how certain pieces of music calm us. Some studies have acknowledged the healing effect of certain types of music or their influence on the growth of plants. What is happening here, really, is that human beings or plants are brought back into resonance with the formative plane of energies.

We can illustrate this with a small experiment: take two identical, or proportionally equivalent tuning forks, and strike one. The other, even if it is several dozen meters away, will also start to resonate with the same note. A single tuning fork could make 1,000 or 10,0000 or ten million others resonate. Underlying this fact are immensely powerful scientific laws – which could also be put to highly dangerous use if they were misused. Handling energies requires awakened consciousness and conscience, for they can generate both life and death, as the atom bomb makes clear. In fact, a science of energies already exists and is being put to secret use in realms such as weather control, pluviometry and cloud cover control. These are weapons that several countries already possess, but discussion of them is forbidden in all political domains.

To sum this up simply, we just need to understand that sound-generating harmonies can 'tune' a place to another vibratory source. In other words we can make archetypal forces resonate elsewhere than the place where they are generated. To do so it is necessary to reproduce the particular angles, proportions, and much else, that correspond to these celestial or planetary sources of powerful vibration. It is not a matter – as has sometimes been done, at Chartres for instance – of simply recording the micro-frequencies at a particular location and then reproducing these. In this case we would merely have a copy, without therapeutic effects.

Activating the energy

To makes these proportions vibrate, finally, in the same way that one makes the strings of a musical instrument vibrate by plucking it, such buildings need to be 'fed' with something that stirs their vibratory forces into motion and makes their energies resound. To do this requires one or several forces. This is where particular properties or 'anomalies' of places chosen by the ancients (electromagnetism, water etc.) enter into play. The human being himself, standing above the mineral, vegetable and animal realms, is of course also a key force in catalyzing such forces. Powerful places attract great individuals, people who are ahead of their time. Something similar is also true in viticulture, where a few great names, whether known or unknown, once set the course of progress. It is a terrible thing to see some wonderful vineyards ravaged in the modern era by lack of understanding or conscience.

The human voice, our thoughts (which also, though we have forgotten this, have a vibratory effect) and our gestures, also form an element that can activate these great buildings, a little like the first violin in an orchestra who 'leads' it. Steiner stated that in Greek architecture the forces are active by themselves, but in Gothic architecture the role of the human being is indispensable for 'unifying' or 'potentizing' the energies expressed there. Thus a place resonates in tune with its 'mothering forces'; to which, to use modern concepts, we can say that it has been 'plugged in.' Let me repeat that before reaching the physical plane everything consists of information-bearing energies borne on all sorts of measurable frequencies and waves. This is the principle, and the architecture of the past knew how to draw on it. Ultimately the growth of a plant, an animal or a person is fairly similar, but in this case all is accomplished via an almost invisible architecture, composed of a system of organized forces that reside in the earth and the atmosphere.

This long explanation was necessary since biodynamic agriculture does something similar, though in a different way. It helps tune a particular location to the macrocosmic laws of life without which the earth would be dead matter.

Water

Water is the essential element for properly establishing such resonance in a living organism, or transmitting all these vibratory life-forces into it. No life without water, as the saying goes. To learn at school that H_2O is water without learning about the underlying forces that unite its two elements is already a first step in denying a knowledge of life forces, and in ensuring that winegrowers later resort to chemicals without much thought. We have to understand how water fulfils its qualitative role as resonator. The molecular connections of water in living creatures are modified by electromagnetic fields and, more generally, by the energy forces in which they are immersed. Practitioners of homeopathy know that water retains a 'memory.' Water is the bearer in the physical world of both terrestrial and cosmic vibrations without which we could not live. The speed of its motion underground, the particular electro-tellurism and magnetism of the place it is passing through, the local geology and even its shapes and configurations – for all these things are connected – can give water a 'range of frequencies,' as we say nowadays, that is very similar to those which prevail in one or other of our organs. Thus so-called healing springs have been known for their beneficial effects on liver, kidneys or heart etc., giving back to a weakened or exhausted organ the life rhythms which it needs. In other words such water reconnects an organ to its archetypal matrix of forces. Every change in frequencies, in the waves which surround and penetrate the earth, change life's expression and thus also change the human being – and can also exhaust his health. The subject is taboo and is never mentioned, for costs associated with *energy pollution* and its effects on health would bankrupt governments. Such pollution includes all the electro-smog and bombardment from technological devices that interferes with natural and cosmic rhythms. Such frequencies are not benevolent but highly destructive, such as high-tension cables, mobile phone antennae etc. These dangerous frequencies are apparently being used to develop a new generation of secret weapons, while the healing ones – for example in alternative and homeopathic medicines – are increasingly outlawed in France. We are now faced with a situation in which man-made appliances are interfering with the natural, life-generating frequencies of the solar system, thus having a detrimental effect on life forces. This also helps explain why the bar codes imposed by law on *all* our food and even our medicines, are so harmful. This is due less to the energies used in barcoding than the rhythm imposed, based on a threefold repetition of the figure 6, giving an energy version of the 666 numeral. We can see visible signs of the weakening caused in the crystallization imaging process (see Plate 15).

Artificial interference

Yes, indeed, the situation today is a good deal more complex than it used to be, for the human being is creating and imposing on our earth, and the whole environment, all sorts of artificial frequencies without considering the interference this represents to the frequencies which sustain life! Solar energies do not arrive on earth via lorries and laborers but through multiple frequencies, each of which has precise functions. To fill, or rather saturate the atmosphere (radar, satellites, antennas etc.) with frequencies created for our electrical goods will come to modify what Kepler called the 'music of the spheres' – that is, all the system of harmonious energies which gives life to the earth. Failing to understand this is, at the same time, to alter the climate and the human being. We must see the earth not as an isolated entity but as embedded in a dynamic macrocosm and interpenetrating systems of forces.

When planet earth is exhausted or reaches the end of a cycle, it recharges itself by inverting its polarities. More concretely, this means that the North pole reverses and becomes the South pole. This is sometimes referred to as a 'pole-shift.' Scientists know that it has already happened somewhere between 150 and 300 times in history. (By drilling in soil with iron magnetite it is possible to discover, depending on the soil's depth, whether the North or South pole was dominant at any time.) It is possible for the lengthy cycle to be shortened by mankind's activities, particularly in relation to our effect on the energies in the atmosphere. In this respect it is important to go beyond the notion of 'global warming,' and see that we are facing the first steps of a pole-shift. We do not know how quickly this inversion can take place. It may take a century or it may happen suddenly in a few days, but will affect all living organization (climatology, ocean currents, species adaptation etc). Migratory patterns have already altered, and, for instance, parrot fish from Senegal are being found off the coast of Italy. Air pilots are finding considerable magnetic changes. In certain instances, magnetic and telluric currents necessary for health have been substantially altered in large urban areas and their surroundings.

This is not science fiction, but a real threat studied seriously by some scientists. When studying the earth, one should consider that it is a living organism. This calls for a unified, holistic thinking – one that can combine various disciplines and data and reveal the truth that lies behind them. In the same way, it is important to see that an earthquake is not, as we are often told, simply caused by two tectonic plates that touch – just as a smile is not just two lips stretching. What is significant is to know the person who stretches his lips into a smile. In other words, it is vital to know the system that under-lies an earthquake – knowledge available only through studying the macrocosmic solar system and its constituents (the planets), whose trajectories continually affect life on earth.

Steiner said once that the earth is like an elephant. For years it can put up with being treated poorly, and then sud-denly it might break its chains.

Nowadays, too many people are captives of the dogma which says that life is born from matter. In fact the reverse is true. It is not the soil which makes the plant but, first and fore-most, the plant which makes the soil and humus by condens-ing the intangible. This lack of understanding for the world of energies and the innumerable waves and frequencies of which it is composed, may become perilous in forthcoming years, and give rise to legal proceedings that dwarf such things as asbestos poisoning or contaminated blood supplies. Can one place one's ear to one of the frequencies of 900 or 1800 million vibrations per second without thereby modifying the whole behavior of one's brain cells – which also have their rhythms and their susceptibilities? Existing studies on mobile phones (cell phones) have been suspended, and the budgets supposedly allocated for their review have been linked to a change of protocol, in order to demonstrate that they are non-repeatable – and can therefore be ignored, provisionally at least.[16]

This is the nature of our times. Did you know that no insurance company will cover possible injury from a mobile?

Specific forms and energies

Thus forms, proportions and the energies to which they are connected, and materials themselves, act upon molecular links. This is why, in the past, each form had a precise function. One may think here of ancient porcelain vessels whose shape is echoed in the wine carafe. Forms affect taste, either 'opening' or 'closing' a food. The same is true in cooking. In the distant past people knew how copper pots could be hand-wrought in a way that exerted powerful effects on the food

---

[16] See studies by Dr. Roger Santini at: www.next-up.org

cooked in it. Materialists would say this is pure imagination, but scientists today can tell you that faceting of the copper changes cubic molecular structure into a pentagonal mesh, thus exerting a different effect on food. We are rediscovering the Platonic forms and their properties. Science is starting to connect with a qualitative world secretly active in our daily life. A team of cancer researchers (as published in Le Monde in May 2007) has discovered 'that the cancer process does not solely result from changes to the DNA but that *it is conditioned by the environment in which this DNA exists*...a cancerous lesion thus appears when DNA and RNA no longer respect universal rules of the genetic code.' This is a remarkable advance.

Even Pasteur experimented with different-shaped containers for conserving water. In certain ones, water retained its freshness for much longer, in others it quickly went stale. Different forces were being invoked here. We cannot speak of the shapes of amphorae, barrels and wine carafes without understanding the energies they invoke. We may be able to better understand this when we look at the fine structures of snow crystals and their infinite, subtle differences. Examining a snow flake we may feel almost compelled to focus on what one might call the 'energy matrix' that has created it – i.e. the extraordinary patterning power of counter- or formative forces which in so many places and with such elegance and precision oppose the force of matter.

The crystal and its counter-image, i.e. the shape of formative counter-space.

This is why certain oriental people, when they observe a landscape, are more interested in observing surrounding space than the solid forms of matter. In doing so they try to perceive the forces that have sculpted the forms of the landscape accessible to their senses. Such observation can allow us to connect our grasp of forms with the different energies related to them.

Below we will examine the effect of shapes – particularly those of barrels – on wine, as well as the influence of specific locations. Ideally we would teach children to sense the energies in a place in the same way that we ourselves have learned to taste wine. If we develop this sensitivity, we can experience within us the effects that oppose gravity, or those which weigh upon us (remembering that we too have our own 'architecture,' e.g. in the arch of our feet and dome of the skull). Likewise we can try to sense how such influences affect our creativity or psychological state, or even how the food we eat resonates within us.

Coming back to shape, it will be clear by now that each shape or form generates specific frequencies or microfrequencies and wave lengths, a knowledge which architecture of the past made full use of. As we have seen, domes, vaults and other forms used very specific proportions and rules to generate forces beneficial to the human being. Everything on earth is subject to underlying numerical laws.

Effects of this knowledge for viticulture

These qualities of invisible, but very real energies also affect our wine. Too much time spent in a place where one does not feel well can lead to severe illnesses. The same applies to a cellar in an 'inappropriate' place which can exert a negative effect on the wine – where the laws of life succumb to an excess of telluric forces. This occurs sometimes in very small areas of a few square meters, where it always proves necessary to maintain fermentation with the aid of a neighboring barrel, or rack it (draw it off) to regain a fresher taste. In contrast there are some ancient cellars, imbued with the wisdom and knowledge of long-gone winegrowers, which can 'swallow' or reharmonize the imbalances in the must without anyone having to do anything. Some people still have this inborn 'gift' and can modify or improve wines by the manner in which they approach them. Others are able to heal or cleanse the energies of a particular place that has been ignorantly harmed. Every viticulturist aware of the laws of life will have interesting stories about such things.

Electric pollution in wine cellars

Many of the finest modern cellars, which sometimes represent huge investments, are quite absurd in terms of energies. Simply on the subject of electrical pollution – i.e. at 50/60 hertz, which emits 50 or 60 vibrations per second to everything it comes in contact with, including cement, from an electric wire or even an unlit bulb – one finds situations which really make one want to shake the architect or the owner! While the electric part of the electromagnetic field can be easily resolved through a good earthing system (which usually deteriorates within a few years) its magnetic part, which is more perverse, is sometimes impossible to cure. In former times people only put low voltage cables (25 volts) or continuous current in cellars. *Poor yeasts, summon your courage to battle with all these obstacles!*

Dear wine lovers, please reflect on your own wine cellars too – I'm thinking in particular of certain bottle fridges which too often emit 50/60 magnetic hertz! And please, above all, do not buy one of those electric field detectors since their sensitivity is sometimes greatly reduced in order to calm your suspicions. Also avoid too much metal in a wine cellar which, together with today's hertz saturation, sets up resonances with all sorts of rhythms injurious to our real, living wines. Such effects may greatly modify their ageing properties...

Wine barrels

The barrel is in fact a permanent arch: just climb into one, or if you're too big, into a 600-liter vat, where it will be more comfortable to experience a few happy moments.

Solar forces of attraction,
forces of levity

Warmth: fruit-forming energy. The degree
of connection with heat-forces leads to a
varying fruit-forming capacity

Light: flowering energy. The degree of
connection with light-forces leads to a
varying flowering capacity

Liquid: leaf energy. The degree of
connection with water-forces leads to
varying leaf sizes

Mineral: root energy. The degree of
connection with gravitational-forces leads
to a varying capacity for penetrating the soil

Forces of terrestrial attraction,
forces of gravity

Plate 1: The Four states of matter

Appollonian Plants

Fine cypress drawn Uppwards

Cedar: long horizontal branches

Willow: related to light & water

Dyonisian Plants

Vine: root forces decend strongly

Plate 2: Four different types of plants linked to the four different forces, in polarity between terrestrial & solar attraction

Plate 3: The nettle concentrates 'heat' and 'light' forces in its leaves.

Plate 4: Through its very strong connection with heat-forces, the cypress tree reveals a wholly upward-rising growth formation, which does not permit any branch to extend horizontally. This is a perfect illustration of what the Greeks called an 'Apollonian' plant.

Downward forces stronger   Upward forces stronger

Plate 5: Vine, laurel

Plate 6: Roots from a vine that was given weed killer after roughly 10-12 years. One can clearly see how its roots then had to grow towards the surface of the soil to feed on fertilizers.

Plate 7:

Triangle

Trefoil

Nettle

Square

Cruciferae

Hydrangea

Pentagon

Periwinkle

Vine

Apple

Plate 8:

Hexagon

Bee cells

Iris

Snow crystal

Octagon

Strawberry

Decagon

Red poppy

Plate 9: A church spire

Horse: Linked to heat & formation of fruit & flavor

Goat: Connected to warmth & light, it survives in hot dry climate. Its forces are oriented towards fruit & flowering

Cow: Water forces are dominant, which correspond to leaf formation

Pig: Earth forces (gravity) is dominant, often feeding on roots. Its forces correspond to root growth

Plate 10: Discovering the dominant state of matter (temperament) in animals in relation to manure suitability for vine growth

Plate 11: Stirring the biodynamic preparations: creating a vortex

The rose showing its spiral form

Spiral in a priest's miter

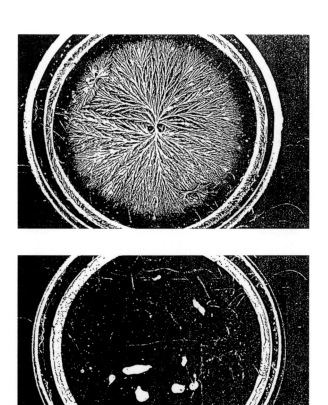

Plate 13: Sugar crystallization images: unrefined (top) and white. The image shows clearly that the white sugar is a 'dead' product. The copper chloride process has not been organized by life forces, but instead diminishes into a dried up heap.

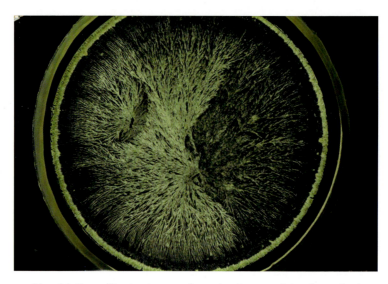

Plate 14: Crystallization images of strawberries: a traditional type (top), and a new variety grown using organic methods. The bottom photo indicates, despite organic farming, the new type of strawberry has poor life-force organization, with no structure or center. The bio-engineering involved in its development indicates negative effects at the enregy level.

White wine: showing excellent structure and detail, as well as a precise center

The same wine after passing under a barcode reader. The pattern now shows distinct areas of weakness

Poor crystallization image of a wine produced using modern farming methods and artificial techniques in the cellar

The importance of dates for harvesting and planting: identical vines, cut and planted in the same soil at an interval of several days

Lettuces planted on either light (blossom) or water (leaf) days, at an interval of several days. Those where a 'light' impulse (left) has reinforced the flowering process have started bolting much earlier (the stem is thick and growing strongly upwards, so these lettuces are no longer fit for sale). The other batch, sown under a water (leaf) impulse, have rounded out nicely and have no main stem inside them

Diogenes wasn't crazy![17] An animal less imprisoned by its intellect than us knows this instinctively. A dog will eagerly exchange its basket for a barrel. Birds do not construct square nests. Their eggs are not cuboid, nor are the honeycomb cells of bees. When we start to observe the animal world and the forms which it generates, we discover all around us, again, the effect of the Platonic forms: the hexagon for the honeycomb, the pentagon for the starfish, and triangles for more primitive forms of life. Each species connects with a certain form or certain forces – which amounts to the same thing. This also shows why so many medicines are derived from the animal kingdom.

Snow crystal

Which form best suits wine? What force does it most need as it develops? Where would we place it in the chain of life? Each person can seek his own answers to such questions. Before entering into these specifics, though, we need first to take a more general, global view as people used to do. The pronounced slope of certain roofs of old houses (54°, which is the also the angle found in a five-pointed star) can lead us to the pentagonal aspect of the five lobes of a vine leaf. Here the Platonic forms are again apparent.

---

[17] Diogenes of Sinope (412-323 BC), Greek philosopher known as 'the Cynic.' He once famously took a tub (depicted as a barrel by John William Waterhouse's painting) for an abode.

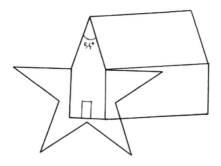

Drawing of roof-slope contained in five-pointed star

One can also understand the shape of certain church roofs in this way which, inverted, give us the form of an amphora – in which wine is very much at home, although in a different way from a barrel. But to properly play its role the amphora ought to be buried in the ground, and thus impregnated with the earth's electromagnetic forces. How else can its forces be received? This knowledge is still alive in the country of Georgia. Calling on the world below or the world above invokes different laws, and exerts a quite different effect on the wine.

We need to rediscover this knowledge to avoid constructing both buildings and containers in any arbitrary way and with any arbitrary material. Each person's creativity should first of all be nourished through knowledge of the laws to which the earth and the human being are subject. Without this, creation risks becoming savage, devoid of meaning and cold, and may in consequence act very negatively on the human being.

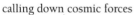

calling down cosmic forces

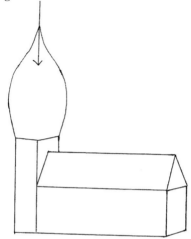

calling up terrestrial forces

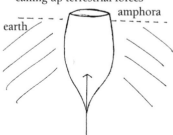

amphora

earth

Church with a roof identical to an amphora. In the church, terrestrial forces are drawn upwards, giving us a sense of elevation. In the amphora, cosmic forces are drawn downwards thus giving rise to a process of incarnation (penetration of physical world).

We also need to regain an understanding of the sun and its effects at different places on the earth, and of the positions where it rises and sets at particular moments, i.e. the summer and winter solstice, or, one can say, the day when solar forces triumph over the laws of gravity and vice versa. This may help us to comprehend the shape of the foundations of house locations up to two or three centuries ago, which varied according to latitude. The further south we go the more we find a long rectangle (which ends with a line at right angles to the equator). The further north we go, in contrast, the more the rectangle becomes first a square and then a north-south orientated rectangle. Why mention this here? Because one finds very similar laws governing different forms given in the past to barrels, depending on the latitude of their provenance. At Porto, for example, the barrels are longer than in Burgundy, for these reasons. Thus during its genesis wine was in harmony with the real energies at work in a particular location.

In the past, as we saw, this knowledge of energies was not just understood but intuitively experienced. There was no need to explain it. One can probably also extend it to different forms given to vessels containing milk, oil and water, which no doubt connected with the body of energies of each product and improved their conservation.

Today we urgently need to reconnect to such knowledge and thus liberate a large part of the scientific world from the cul-de-sac in which its materialistic dogmas have imprisoned it.

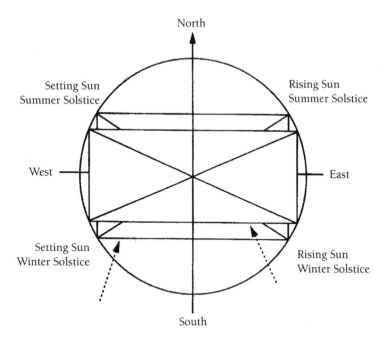

The four angles of these shapes are linked to the rising and setting of the sun at a specific latitude.

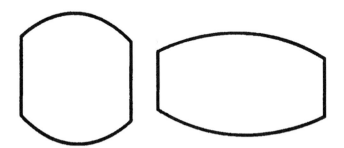

Traditional barrel shapes: the further south one travels (in the Northern Hemisphere) the longer the barrel's shape.

Variation of forms in relation to latitude [18]

---

[18] In: Eglise romane, J. Bonvin and R. Montercy, Editions Mosaique

It is time to place these countless and respectable discoveries gained through laborious investigation of matter into a wider context in which the physical aspect is just part of the reality.

We should add here that giving an animal poor nourishment, overriding its reliable instincts with artificial aromas by making it eat 'molecular compounds' that do not suit it, in fact isolates it from connection with formative forces. In the domain of health everything relates to universals.

Globality

As we saw in Chapter 1, being sundered from a universal context found expression in ancient times in the myth of Dionysus. The Egyptians also referred to this terrestrial state in speaking of Isis who was desolate as she searched everywhere on earth for Osiris who had been torn to pieces by Typhon. Typhon symbolizes gravity. Here again we have an image of incarnation into matter, and our separation from an overall context of energies through gravitational laws. Isis, one can say, tries to find on earth each small piece of a puzzle which, at the level of energies previously formed a whole, beautiful image named Osiris. This puzzle, fiendishly difficult to resolve, relates very much to the search in agriculture and viticulture for ways to produce a true and great wine.

Material science cannot advance unless its knowledge is bedded back into a much wider context than the limits within which it lies fragmented today. Without this, indeed, it will become dramatically dangerous, as genetic engineering has already shown us, with its tendency to believe that the 'bad' gene is responsible for lack of equilibrium which has arisen by 'unlucky chance.' No one tries to understand why the system of energies acts in a different way in one case or another. Instead people hold fast to the fact that, to promote what is best, they need to suppress what is unwelcome. Nor is any consideration given to the reaction of an animal that responds immediately to genetic engineering by developing disease such as leukemia or, if you prefer, to a disorganizing influence with a disorganizing reaction. On the contrary, attempts are even being made to prevent leukemia through genetics.

And yet we will have to learn, with suffering and sorrow perhaps, that the only thing enabling us to avoid the terrible degeneracy of our immense materialism will be knowledge of complementarities linked to a more global definition of the laws of earthly life. It is this understanding alone that will bring about true and lasting progress in agriculture, in such a way that the laws of the physical world can be connected to the surrounding world of energies. This path already lies open and it is called biodynamics. Each day it reveals its results in wine a little more, and will bring about a different approach in agriculture since its qualitative effects are undeniable, proven and without any injurious side effects. Many influential magazines have reported on organic viticulture, but fail to discuss the numerous arbitrary and artificial tastes that are added, which can be avoided through biodynamic wine growing.

Biodynamic agriculture is extraordinarily economical, especially if one factors in the enormous costs for social benefit payments linked to dietary deficiencies. And it is these last arguments which, for commercial reasons, have made it an enemy to beat down at any price. It has been called sectarian, something clearly refuted in legal proceedings (and by courageous journalists). Then it has been declared something 'for the wealthy' despite the fact that it is employed by many small viticulturists and in developing countries. It has been called 'esoteric,' since it works with very tiny material quantities. If biodynamics is esoteric, however, because one cannot explain its effects in purely physical terms, then a mobile phone or radar which enables an airplane to find its destination from thousands of kilometers away can be called the same.

Some people in INRA (French National Institute for Agricultural Research) who are dogged in their criticism despite the enormous damage and debts they continue to inflict on both human beings and the earth, nowadays call it 'obscurantist' without wishing to understand that their instruments simply cannot measure these energy phenomena – which they therefore refuse to believe exist despite the convincing qualitative results.

The effects of biodynamics derive from real laws, just as much as the law Newton discovered when watching his apple fall. It is a shame, in fact, that people have focused on this law of gravity so exclusively, for Newton also spoke of opposing principles – of solar laws which must be perceived in a different way than by instruments geared solely to the world of matter. And this is something which is once more coming to light thanks to certain physicists, sometimes much scorned by their colleagues, who discover that behind the physical world there is another, which is wholly overlooked. The void is not the void, they say, but is full of other forces. Science is taking a great step forward here.

For us viticulturists and for all who are passionate about wine, what counts above all are the results; and as far as I am aware no viticulturist using conventional agriculture can, without artificial aids and technologies, equal the quality obtained by biodynamic methods, this agriculture of the future, nor even approach the originality of these wines when things are properly managed. Thanks to research into the underlying forces that sustain life we can ask ourselves the right questions about our vines and our farms too. What do they need to unfold their full expression? Which animals are needed, or what landscape? What kinds of biodynamic treatments, vine stocks, slope and directions? In other words, which forces are active in all these criteria?

We have spoken of frequencies, of a world of energies underlying and sustaining life, of archetypal forces. All these little understood words seem a little primitive at the first reading. What is biodynamics really? What are the actual measures we use for our vines? How is it different from other methods of cultivation? It is time to get to grips with this subject.

4

Biodynamics in Viticulture

'Bio' means life; and 'dynamic,' or 'dynamism' is an enhancement, acceleration or also a stimulation of life. How does one dynamize life? It's very simple really. Let us say first of all that these processes already exist in nature, partially at least. Everyone knows, for example, that a mineral water, before being immobilized in a bottle, is much more biologically active and dynamic because of the movements to which it has been subject while circulating underground. One can also say that rainwater, which everyone knows is much better for plant growth than tap water, has been dynamized by the atmosphere.

Biodynamic agriculture began in 1924 in Germany, at the estate of Count and Countess Keyserlingk in Koberwitz, when Rudolf Steiner decided to give some very precise suggestions to improve the health of plants, their taste and their nutritional qualities. He gave advice for developing some specific natural preparations to enhance life processes, which plants need to express themselves fully. To do this he first selected some medicinal plants.

These were chamomile, yarrow, nettle, oak bark, dandelion and valerian. Then, for some of these, he identified corresponding animal organs, with which they create a synergy. He advised that most plants should be inserted into these animal organs for a certain period to reinforce their effects and, usually in winter – though in summer in one case – should be buried in the ground under fairly specific conditions, and be dug up again in spring.

The preparations

Let us explain this in very clear terms to avoid the accusation of sorcery by our detractors – who are becoming ever more virulently opposed to these methods in the face of the qualitative success of biodynamic viticulture in increasing numbers of countries. Let us take a specific example: everyone knows that chamomile has a connection with or particular effect on the digestion, and more specifically on the intestines. We also know that animals are more evolved and therefore higher than plants. Plants are wholly subject to the living organism surrounding them, composed of the earth, the sun, the climate etc., which they are wholly dependent on. The animal, on the other hand, is more developed and internalizes the world in different organs: respiratory, digestive, cerebral etc., which give it a certain autonomy towards the external world that allows it to drink, eat or sleep where it wants. We already see this direction in cellular reproduction accomplished by invagination, somewhat different from cellular multiplication in the plant kingdom. Certain plants – the orchids in particular – take on slightly animal characteristics, those of insects for example, revealing perhaps the first tentative step towards developing this superior state.

To understand this idea better one needs to read the books by Pelikan,[19] referred to previously, on medicinal plants. It is always the case that this superior level of the animal world has powerful effects on plant growth. Everyone knows of course that animal manure is best for promoting plant growth. And what is this manure? Simply vegetable matter that has passed through an animal's digestive system. Except of course when crazy people, unable to distinguish the qualitative difference between a vegetable and animal protein, advise feeding cows with meat, which can make them go mad ('mad cow disease').

One can say, therefore, that manure is vegetable matter impregnated with an animal's metabolic forces. We all know, as well, that medicine often has recourse to remedies derived from the animal kingdom. These explanations – which are given here in very simplified form – can help us nevertheless to better understand the synergy that can exist between an animal organ and a plant, and provide the basis for developing certain natural preparations advised by Rudolf Steiner. I say 'very simplified' since the choice of plants, animal organs and animals involves profound understanding of archetypal life forces which are concentrated in each one of these preparations (see Plate 10).

---

[19] Wilhelm Pelikan, *Healing Plants* (op. cit.).

Ultimately, by enclosing chamomile in an intestine, its properties will be reinforced or enhanced. So now all that remains is to choose the animal which has attained the greatest development of its digestive system, or which is the biggest ruminant in Europe. This is the cow, of course, which many ancient civilizations rightly thought of as sacred, something also reflected in Christianity (the ox and ass in the stable).

So one takes a cow's intestine and stuffs it with chamomile, and buries it in winter. When spring comes one digs up this preparation again and adds very small quantities of it to a heap of manure or to a small pile of dung which is subsequently dynamized in water and sprayed on the soil at particular times. This is the principle. Enclosing the plant in an organ thus serves to enhance or activate a specific process. Certain plants are used without the aid of an animal organ (nettle and valerian), while others are inserted into other organs. The organ chosen for yarrow is the bladder of the deer, an animal figuring widely in symbolic or allegorical texts, whose antlers give it immense sensitivity. For the dandelion one uses the cow's mesentery, the fine, silica-rich interior wall enclosing the cow's intestines. Oak bark is inserted into the vessel of a domestic animal's skull. Here we see the vegetable calcium of oak bark (77% calcium content) placed in synergy with the calcium in the skull of an animal.

All this is easily understood. There are much more profound explanations, though these too are no doubt incomplete, but there is no space to go further into this here. The sole aim of this book is to help wine lovers to understand why biodynamics has real effects which improve taste, longevity etc. When we speak of the specificity of each organ and of each animal used for these preparations, and include the forces which underlie life, everything becomes much more complex. Many justified questions cannot be discussed here, for instance the effect of burying the preparations in winter or summer, of burying them deep or closer to the surface, in a very dry or damp soil, in fertile or muddy ground. Why a cow and not an ox, a deer rather than a pig? Let us say, simply, that nothing is left to chance and that every detail has its importance for achieving precise effects. Those who wish to explore this approach further are referred to the Bibliography.

Thus the biodynamic preparations enable us to use the synergies available in nature, and all we need to do is know how to use them. Each preparation creates a 'place of reception' or 'microbial entrance' for specific forces needed by the plant.

The cow's horn

We still need to mention one, very precious, organ which our detractors try to ridicule: the cow's horn used for two preparations.

This organ is of immense importance for the cow. In stark contrast to the deer's antlers which act like antennae, the horn and hoof serve first and foremost, according to Steiner, to retain in the cow's interior the currents of forces which would otherwise seek to escape. Thus the horn and the hoof act as a kind of internal reflector. Outwardly it participates in the connection which the animal has with solar forces of ascension, something one can also see in ancient Egyptian depictions of the goddess Hathor (who brings the heavenly gift of wine to the earth) with the head of a cow and a solar disc situated between her horns. The horn enables the cow to bear its heavy head with ease and grace. Its horns are permeated with a specific process of great interest to viticulturists who are continually seeking to improve the taste of their wine. It is true, of course, that agricultural schools nowadays advise de-horning cows, a widespread practice which people carry out without awareness of its significance. What use are these horns, they ask.

In ancient times the horn was viewed as the source of richness and abundance. Old paintings depict the 'cornucopia' from which abundance, pieces of gold or other precious objects pour forth, but today few understand neither this symbol nor the forces it represents. In Denmark and Georgia they still sell horns inlaid with silver for drinking wine or water – the last vestige of past times when, to announce that it was time to eat, people still used the phrase 'The water is horned.' Yes, indeed, water was poured into a horn which instilled into it the frequencies the horn retains for several years after its removal, giving the water health-giving properties. In Georgia one can still find beautiful cow horns intended for wine drinking on feast days.

In the case of biodynamic preparations one fills the horns with cow dung and buries them in the winter in a carefully chosen place. If, by way of experiment, one buries next to such horns the same dung in a little terracotta pot one finds, on digging both up again in spring, that the dung in the horn contains about 70 times more bacteriological activity. Yet again we discover that a proper understanding of the life that surrounds us can invoke powerful forces which will benefit our plants. Only the contents of one or two horns are used per hectare, which may come to 100 or 150 grams of dung. How does one spread such little quantities over a whole hectare? It is very simple. Without going into every detail, we dynamize this dung in 60 or 70 liters of lukewarm water. How? By creating a vortex or spiraling eddy in a container of the mixture, either by using a long, suspended stick if one does it by hand, or by using a dynamising machine. In less than a minute, when this vortex reaches almost to the bottom of the container, one reverses it as quickly as possible, and then continues the process (see Plate 11). This fierce reversal creates what one can call a chaos, which is essential and needs explaining. We spoke above of the figure 8 or lemniscate, which is really just a circle that has been twisted. This twisting motion reverses the direction of rotation of energies in the other half of the 8. Above they move from right to left, while below they turn from left to right. The moment of reversal passes through a point where the two circles of the 8 touch each other. This 'moment' is the indispensable chaos. At this precise point we are no longer in the world of energies nor yet in the physical world. This 'point between two worlds' is that of chaos. It is this which

Celtic spiral

induces the child to cry at birth and releases us suddenly at death. Such momentary chaos is important for bringing down the forces of which biodynamics makes use. Disorder invokes an order subject to new laws, those of the earth. We can also relate this reversal to daily life: some drama and the chaos it creates can sometimes open us up, and allow something else to enter, some radical change to occur.

Returning to our dynamisation, we can say, quite simply, that this strong agitation enables the properties of the horned dung to enter the water mixture that is subsequently sprayed on the soil. We can also say that the latent forces un-

A diffuse, peripheral world concentrates in a point. Here non-matter becomes matter as life incarnates

derlying life often manifest in physical matter in the form of spirals. One can see this in the petals of the rose, a wonderful, opening spiral formation, in the arrangement of seeds in a sunflower, in the position of shoots around the stem of certain plants including the rose, and also to some extent in the tufts of animal fur or the hairs on a human head. This form is engraved on some menhirs, manifests of course in cyclones, and on shells etc., and can be found everywhere around us if we are attentive to it (see plate 12).

Studies are currently being carried out on spirals by the CNRS (French National Council for Scientific Research). The book by Theodor Schwenk, *Sensitive Chaos*, shows many examples and illustrations of this. What is the significance of this spiral form? In simple terms, a macrocosm that becomes a microcosm; a periphery that becomes a center; a journey from something dispersed and intangible to the condensation point of matter. In the northern hemisphere, the spiral invokes incarnation (penetration of matter) when it rotates clockwise, and excarnation (departure from matter) in an anti-clockwise direction. Many wine lovers know this instinctively, often rotating the wine in their glass from right to left as though to better extract its aroma. In principle, this direction would be reversed in the southern hemisphere, the clockwise rotation there invoking ascent.

Using biodynamic preparations, and their properties

The preparation made by these means is sprayed on the soil in the evening as the sinking sun draws the atmosphere towards the earth, condensing it into dew.

To understand how it works we just need to realize that each drop of this preparation is a vector or mediator of microbial life that can activate our soil and help it to develop the mycorhiza we mentioned earlier. You will recall that the more intense microbial life is the better vine roots can seize hold through them of all geological aspects of the soil. Thus we introduce into the soil life processes starting from micro-organisms of all kinds which have been activated by passing through a horn. Basically the horn acts as a nursery that cultivates the life of micro-organisms in the dung we put into it. Each one is the mediator of, often different, information. Then, through dynamisation, one inseminates the soil with all these various processes. Thus we imbue the soil with life processes which can only develop if the ground is 'receptive' or, if you like, where there is a welcoming milieu. For this to work it is necessary for the soil to be free of these terrible poisons which are increasingly injuring human health but whose use is still advised in agriculture in the form of herbicides and chemical treatments. One cannot engender a life process on the one hand, and on the other initiate a death process and think this will achieve something! This remains true even where people nowadays try to use less toxic substances or less of them (so-called conservation agriculture). These substances still strike 'false notes' or if you prefer act as blockages to the full expression of a system which the world of science overlooks. We need to make a choice for one or the other approach, and not mix the two.

The highly fertilizing property of horns has been known since the beginning of time. They are sold in the form of bone-meal or powder by almost all agricultural associations.

Let me mention in passing that the compulsory treatment of this product, recently introduced because of a continued failure to understand mad cow syndrome, has rendered it much less effective. It is interesting to note that in France, whenever a natural product has a significant effect, a pretext is found – always in the name of people's supposed welfare – to diminish its qualities. We have seen the same thing recently in relation to homeopathic substances used since the dawn of time, which are still accepted in other countries.

Ultimately the sole difference, though an important one, between ground horns and this biodynamic preparation is that in biodynamics we use the horn as a container for the dung, to concentrate its forces, rather than as matter itself. The dung placed in the horn is impregnated during the winter with forces or dynamisms active in both the horn and the earth, which enable the horn to be preserved for three to four years.

The horn serves to develop another preparation too. We fill it with quartz instead of dung – that is, silica in the form of very fine powder. Our earth contains this in great quantities and Steiner tells us that it plays an important role in maintaining equilibrium in nature, even if it appears to be inert. It acts as link or medium, a sort of receptive antenna, to certain forces which radiate from the solar system.

To make this preparation one fills the horn with quartz powder and buries it in the ground during summer. This powder, subsequently removed from the horn again, is imbued with very particular energies, frequencies and information. The word 'information' may alarm some people, and yet everyone knows that quartz or silica is used in information technology as a medium of information, or alternatively can be 'charged' to make our watches or other equipment function. In our particular case the charge or the information is given by nature via the horn and the earth in which it is buried, this time during summer as the preparation is destined to stimulate the leaves and photosynthesis. This is very simple isn't it?

To use this preparation we take infinitesimal quantities of a few grams per hectare. In speaking of frequencies weight has no significance – this would make no sense. In biodynamics we are not dealing with quantities but, as we have seen, with energies. You are not charged for your mobile phone use by the weight of its waves – say 2 grams for Paris and 100 grams for New York! You do not receive your TV programs by weight, but nevertheless you see very definite visual effects. When you turn the button on your radio you locate different programs by turning to different frequencies – and all of this seems normal because you are used to it. But it is no less true to say that the descent of life to earth, whose material effects become visible in spring, obeys somewhat similar laws. This means that trying to understand the life of a plant by focusing exclusively on the plant itself, as modern science too often does, is as useless as looking for the presenter of a TV program inside the television itself. The plant and the seed are but a receiver, linked to a vast system we need to understand if we are going to make use of it.

We have been told that genes are merely intermediaries which obey the orders they receive. Biodynamic preparations function somewhat like transmitters, mobilizing and activating precise energies and processes which plants nowadays receive to a much lesser extent due both to the disruption people have caused in the soil and also in the atmosphere in the form of radio waves etc. Until now people have regarded the atmosphere as an energy dustbin into which one can dump whatever frequencies one likes, high or low, without even suspecting that this has direct effects on life. No serious funding is made available for conclusions which might have such a seriously anti-commercial impact, and even if it was, very well-organized lobbies would ensure that research does not go down a path that could lead to 'wrong conclusions.' But we have a responsibility to future generations, and it is high time we became aware of the effects of our actions, and found the courage to put long-term health before short-term profit.

To recap, then, each biodynamic preparation is the bearer of a specific process at the energy level. It acts somewhat like a sensor or vector. The brilliant thing about biodynamic agriculture is that it mediates information to and from this invisible but very real system of energies which is activated to give life to the earth and to each plant. It stimulates and directs it. It acts at a stage before life becomes, or is imprisoned in matter. All this will come to seem self-evident in a few decades.

Thus invisible energy information is mediated by the quartz via the horn and the summer earth. Then we put a few grams of the silica into water and dynamize it for an hour, and then spray it in the morning on the leaves rather than the earth itself, for otherwise it will not have the desired effect. This needs to happen early in the morning because this is the moment when the rising sun acts like the spring season, drawing sap upwards and connecting the plant to upper worlds. What are we really doing when we use this preparation? We are, in fact, using the quartz to draw in the air's luminosity. In today's sadly mundane terms we might say that we are 'passing light information' and thus activating photosynthesis. Also, if the moment is well-chosen, this strengthens a connection with the forces that give rise to taste. The preparation is extremely powerful. It can also help the vine to exhale excess water or act as anti-fungal treatment. Without entering into all the details, the main thing to remember is that it needs to be used with care and caution since today's atmosphere is no longer the same as it was in 1924 when Steiner first gave his indications in the Agriculture Course at the Koberwitz estate of Count and Countess Keyserlingk. Today's atmosphere is less alive. Steiner spoke of the atmosphere as the place where cosmic laws tune themselves to earthly laws – something which is less effective nowadays. If the weather is too dry or hot, this treatment should not be given.[20]

---

[20] See Nicolas Joly, *Wine from Sky to Earth* (Acres U.S.A., Texas), which gives more detailed advice for practising farmers and viticulturists.

You now have enough insight into biodynamics to understand it in principle, and to grasp how it acts or, if you prefer, how it connects the vine to processes working in nature since time immemorial. Biodynamics seeks only to activate. It does not set up frequency fields in the atmosphere to use them subsequently, as mobile phones do, but it makes use of what nature has already put in place to generate life on earth and merely activates and enhances this.

To sum up, then, very small quantities of plant-based preparations are placed into a manure heap or into a few kilos of dung, and act as catalysts for processes which the plant depends on to express itself fully in the physical realm: processes of potassium, calcium, iron, silica, phosphorus; and also of fruiting – in other words, very briefly, of what halts the growth of leaves and branches and enables fruit to develop. One can even understand these preparations by saying that by inserting them into a manure heap they will act a little like the working of organs in an animal, each one having a very precise role to play. Specifically, via micro-fauna that are different and very particular in each case, they will reinforce very precise functions in the soil and the plant. If they have been inserted into manure this will be spread in the autumn, preferably digging it in to the soil. If, instead, they are inserted into small quantities of dung, this will be dynamized and then applied to the soil. The autumn is the ideal moment for these tasks, for then the sun that has become ascendant in the southern hemisphere draws everything downwards in the northern hemisphere, and the soil becomes ready to 'receive' or welcome to its bosom the impulses one desires to give it.[21] Later, in the midst of winter, when cold reigns, the earth crystallizes and, through its crystals, hearkens to a more distant world. We might say that this is the time when the earth, via its crystals, is 'inspired' by the solar and stellar system – an inspiration which will come to full creative expression as spring returns and plants begin to reach up again to their macrocosmic source, producing at the same time sustenance for animals and human beings. Mother earth is continually inspired by the solar and stellar system. During winter she is less active, although not so much resting as receiving influences. This winter moment is very precious for all that occurs beyond the physical plane seemingly immersed in sleep. This is the moment when life, unleashed to some extent from its connection with matter, is present only in the form of energy. Steiner tells us that this is the time of year when we can inwardly hearken to our fields and our vines, harmonizing them through our thoughts alone. This was very common in the past, when farmers would 'keep the midnight watch' or gather at one neighbor's house or another in turn, to reflect, listen, think and understand things better than the intellect alone allowed. If we deepen these suggestions of a somewhat meditative nature, we can discover here a rich realm of almost

---

[21] In the southern hemisphere, of course, this applies in reverse.

untapped possibilities which we can try to develop if we so wish. So-called 'green fingers' are an unconscious version of this faculty. By reflecting in this way the human being can no doubt try to consciously reacquire such capacities.

Particular functions of the cow-horn preparation

The two preparations placed in cow horns have functions that are quite distinct from the other preparations. The one made with dung acts like a growth vector in the soil, mobilizing the life of the soil and connecting it to the vine's roots through an intense microbial life. Thus it helps recreate the vine's physical body, of which, after pruning, there is nothing left but the stock and a few shoots and buds from the previous year. At the beginning of spring, therefore, we need to help the vine to develop as the buds are about to open.

The quartz-based preparation acts via the light and thus has *different effects according to the season when it is used.* The vine, after all, does different things in spring and autumn. This preparation, one can say, accompanies the vine in its daily task. In spring, when the vine is in full growth and creating matter to give birth to its shoots, tendrils and leaves, its help is from above as it were, that is to say through the leaves via photosynthesis. Applied several weeks after the summer solstice, when the days begin to grow shorter again, at the moment when the vine is fully mobilized to mature its grapes, the preparation helps these to form their sugars and tastes via an increased influx of air-borne luminosity. It is this warmth of luminosity, carried on the air, which mediates forces of the solar world that generate taste, scent and aroma.

I trust that it is becoming clear that it is not enough just to 'practice' a biodynamic method in order to obtain a good wine but we also need to try to fully understand how it works. Our understanding of the preparations, of our terrain, the vitality of the vines, the grape variety, the growth forces in the soil, the climatic profile of the year, all help the viticulturist to take the decisions he needs to. Thus one can use both of the preparations made in the horn to achieve a vine that is either more or less present on the physical plane, or, if you like, more or less active in its Dionysian combativeness. At one extreme we can have a large vine with lustrous leaves whose forces of fruiting and taste will be lessened by this exuberance of foliage. At the other extreme we can have a smaller vine with thinner branches, but whose wine will reveal strength and beauty. Exuberance and fruit formation are somewhat opposed. One frequently sees a tendency to one or other of these two extremes.

Giving back creativity to agriculture

Biodynamics is the first type of agriculture, at least for several centuries, to give back to cultivators the possibility of affecting plant behavior on a plane other than the physical – which only allows it limited potential for diversity. This agriculture involves a labor that is almost artistic when we start to ask ourselves how to help a vine best express itself. The possible choices are very diverse. One can reinforce the vine's connections to particular functions to which, at present, it has only poor access. Take the silica process for example – why increase or reduce this? Simply, first of all, because the earth of today is no longer what it was in former times. The earth also grows older, and no longer has such a powerful capacity for growth as it had a few millennia ago, when it developed immense trees – just look at the thickness of the coal strata – or enormous animals. Finally, and above all, our progress in the physical, material realm has often been disastrous for the realm of energies. This aspect is only just starting to become the subject of research but, as I have said several times elsewhere – and without any political emphasis, since ecology really should not be politicized – powerful vested interests try to silence such studies for as long as possible.

The silver lining to the cloud, and there always is one, is that all this gives viticulturists and genuine cultivators the chance to work qualitatively in a way that is increasingly recognized and appreciated by consumers throughout the world. By using biodynamics one can, in the same vineyard, produce a wine that is more or less luminous, more or less earthy, more or less floral, etc. And isn't it also to some extent up to the winegrower, whose situation compared to farmers has been somewhat privileged over the past two decades, to show that all these subtleties worthy of securing an AOC designation can likewise be found in certain milk products, vegetables or fruits? This would encourage consumers to use their purchasing power to make careful choices about the foods they buy, full of original, authentic tastes that earth forces give when one knows how to invoke them. The viticulturist would be able to demonstrate a certain creativity if he wishes, and engage in other actions, with no need to make a secret of them. For example, in adverse climatic conditions, one can apply herbal teas to vines to help them accomplish their task. The vine is a living being. People believe that it fulfils its nature automatically, but this is far from true. Creating sugar in a grape is laborious work, for example. It is possible to use a rose hip tea (fruit of the sweet briar) to stimulate the formation of sugars a few weeks prior to harvest.

One can protect the vine from the impact of sunshine rendered more aggressive by our irresponsible acts, by treating it with a tea prepared from seaweed. Such actions do not violate the vine as technology does. Instead of treating it by purely physical means these teas call forth the vine's own reactions of sympathy and antipathy.

One can also ask which animal is most appropriate to our land, as this too will play something like the role of an organ in the 'living organism' which an agricultural entity represents. Each animal, very directly through its dung, will have a different, tangible effect on the taste of a wine. Each animal is subject in different ways to the four states of matter, and through its dominating principle will affect the roots, the leaves, flowers or fruits. The horse gives a better taste to the wine than the pig, for example. We can try to rediscover these synergies so as to integrate all the constituents of life into a harmonious whole.

Different understanding – different actions

One of the important things which Steiner states about bio-dynamic agriculture (and anthroposophic medicine) is that it does not 'combat' a disease but promotes an equilibrium, and by this means renders the carriers of disease far less potent. This holistic notion of the agricultural organism is an important one. Thus it may be a very good idea to sacrifice a little of our AOC terroir to leave a field, a wood, fallow ground or at least a tree! Every aspect of life, and therefore every species of plant or animal, will attract a different fauna of birds, insects and microbes, etc. which will influence the life of our soils and the mycorhiza of which the vine has such need to unfold the full distinction of its AOC. Every intervention that accords with the overall harmony of the farm or vineyard will not only limit the influx of diseases but will also reinforce the incredibly complex system which gives rise to a wine's beauty, aromas, colors and health-giving properties.

Would there be any danger of avian flu if birds had not been raised in tightly packed battery cages? Has a single state veterinarian had the courage to say so? And if he were to do this, wouldn't he lose his job? Teaching today in the realm of animal health has sunk to the depths of materialism and intellectual aridity. Think of all those young cows gorged on silage – to produce high milk yields – which have to be killed before the age of seven in order to hide their cirrhosis, which would render them unsellable. Think of all the animal proteins which farmers continue to feed them, heating them to 600° to prevent cows – temporarily at least – from passing on mad cow disease. Think of the huge reluctance to understand that animal and vegetable proteins are qualitatively different, and that the cow is not a carnivore. Think of sheep scrapie and swine fever. The list is long and costly, very costly. Should such costs, at least, not motivate us to change our approach? All this ultimately shows us the huge failure of a system of training which has lost its way in holding fast to the microscopically small.

We are plunged so deep in error that we have the right, as taxpayers, to ask this question. We have, surely, a veterinary service which never opposes irresponsible intensive farming and animal rearing practices, and is therefore complicit in error. It is due only to over-use of antibiotics that these almost barbarian agricultural acts have not yet manifested in innumerable consequences for the human being. Veterinary teaching is so narrow that the members of this profession – although fortunately there are always exceptions – sometimes seem like unwitting torturers who focus exclusively on treating symptoms with all sorts of medicines or vaccines with serious secondary effects, without ever trying to address the cause. When will they have the courage, the conscience, to say that 15 chickens squashed together in one square meter are too many? Or that a pig housed in only two square meters is shameful? Why, as veterinarians serving the animal world and the human beings that are fed by it, do they not insist on decent husbandry standards? Soon, when the whole life system has been destroyed, they will no doubt make out that it is only due to them that we are still alive! These are the same people who use a dangerous, internal insecticide treatment just to treat cattle grubs – a simple gadfly that pierces the skin – which you then eat in your meat without realizing it. The unbelievable pretext for this treatment is that these insect bites reduce the value of the hide. In this same case, though, thanks to the courageous intervention of an open-minded minister in the agriculture department, organic farmers have obtained the right to avoid this absurd law, indirectly issued by powerful lobby groups.

The new toy astutely slipped into the hands of these veterinary services is the electronic chip implanted in animals, which imposes on them a rhythm which isn't their own, of course, and which thus distances them from their own rhythmic system, the very core of their health. Really they are implanted with a mini computer. The commercial argument behind this measure, which is still presented as progress, is that it prevents animals being stolen. The truth is that if they are, it is easy to detect this chip and remove it. And secondly, this is yet another very lucrative sales market. Examples of such aberrant measures are too numerous to ignore. The 'Paris Appeal,' a petition submitted to the public services by high-profile oncologists – alarmed by the growing number of children admitted for treatment – and by other well-known individuals, demonstrates a just and effective means for limiting all these abuses.

The same problems rear their heads in viticulture, agriculture and animal rearing: Should one treat the cause or the effect? What is health? In choosing always only to act in response to the effects of absurd practices we will never achieve much progress, that is, grow health-giving foods. Is it necessary to initiate legal proceedings in order to show that the avian flu virus, which has made vast profits for vaccine producers, has existed for a long time, and that birds have carried this virus for decades, only rarely dying from it? It only becomes dangerous, in fact, when rearing conditions are so ridiculous.

Forgive me for this digression from the subject of wine; but to try to avoid such dramas continuing and multiplying with exponential speed, the consumer needs to be warned. One can even regard it as a civic duty to do so. After all, the consumer rules at the end of the day. It is thus right to tell him the full truth. If he changes his purchasing habits the strategies and tricks of big vested interests will crumble in a second. Organic farming should not be immune from this either, for there too one can find chickens in very cramped conditions (sometimes even 6 to a square meter) that are exhausted by the age of nine months.

Finally we need to look at a system to which we have referred above on several occasions, one which brings life and health: I mean the solar system. What does it mean for the earth? Why does the earth belong to a solar system? What would it be without it?

5

The Solar and Stellar System and its Effects on the Earth

In our daily life we are more aware of our connection with the sun than with the integral totality of a solar system in which all the planets and the earth are involved. This awareness of the sun develops most in its absence which we sense, quickly enough, as a cruel lack, whose return rapidly gives us back a feeling of comfort and even indolence. Apart from the sun we are often also attentive to the moon which we admire when it is magnificently full. This is often the extent of many people's awareness of the solar system. The moment we speak of a 'system' in fact, we need to understand that all these constituents mutually interact as in a living organism in which each organ contributes in its own way, and with its own inherent characteristics, to a global whole which is greater than the sum of its parts. It is thus through a kind of synergy that each of the constituents of the solar system, including the planets, affect the earth, and that the earth, in turn, acts upon this system. A team of Scandinavian scientists has found that every nuclear explosion on earth gives rise, approximately 100 days later, to precise signals from the sun in response. This knowledge is something we need to rediscover, and it is encouraging to find that such work is already underway.

To reassure our Cartesians, though, let us ask ourselves first how and by what means such effects might actually exert an influence on the earth. Our first impulse is probably to say that something located millions of kilometers away cannot really affect our physical environment or us in any tangible way. And this is the nub of the problem, where we find ourselves trapped by the limits of the physical, sensory world. Throughout this book I have reiterated that the forces of life do not belong to matter itself but to a complex and organized world of energies, made of wavelengths of all kinds which certain scientists are trying to identify. Isn't it by means of this same world of energies that people can send orders from the earth to a robot standing on Mars, or a satellite of Saturn, which, a few seconds later, carries them out? Likewise, we have already seen that the sun acts upon our earth by means of diverse wavelengths. Aren't we in fact completely connected to it via an overall plane of energies? So why should we deny a similar connection to other parts of the solar system, or to the more distant stellar system called the zodiac? Let us recall once again that our senses only perceive a small part of all that exists.

We also need to be aware that, thanks to certain energy forces, atoms cohere together and enable matter to become accessible to our senses. The cells cohere according to the information they receive. Observing all these transformations in embryology, as mentioned earlier, is a fascinating study.

The whole earth, right into its smallest corners, is continually bombarded by life forces or all sorts of frequencies and cosmic wavelengths whose origin we are unaware of. We know of their existence but not yet their meaning. A big international funding program is even constructing ultra-powerful antennae and satellite dishes to try to decipher the language or coherence of these frequencies. Our sole error is to impose on them our own, very terrestrial interpretation without understanding the underlying energy planes that direct them.

Macroscopic vision

When we approach life in a different way than via infinitely small atoms we enter instead into a quite different kind of understanding, which we can summarize as follows:

Everything that lives is sustained by different rhythms, frequencies, and wavelengths. Physical death can be seen as just the absence of certain rhythms. Each planet, or more precisely each planetary sphere – since each physical planet can be seen as the focus of a sphere of activity rather than its sole, active element – has its own characteristic 'energy language.' The closer we approach to the sun, the more the frequencies mix and interpenetrate which are linked to the concentric circles that the planets and the earth describe around the solar star. I am not going to try to explain here the incredibly creative complexity of this system in its effects on the earth. Steiner, Elisabeth Vreede, Rudolf Hauschka, Walter Cloos, Kolisko, Fritz Julius and many others have all written volumes on this subject which one can study and gradually assimilate. Here I just want to stress, firstly, that before becoming physical all that composes life is in the form of energy information to which matter obediently responds. Matter is never the mould, only the content of the mould, something that is increasingly being understood. Recently researchers have discovered a way of charting the energy maps of each human illness and have recorded them as part of a promising project which may open the doors to very economical preventive methods. Diagnosis of a precise field of resonance, which creates the preconditions for a physical illness to manifest, will enable us to act before the latter occurs. If we discover these frequency anomalies, or this resonance map of an illness, we will know that the danger of this illness appearing is great. This approach can help us understand that energy information or a resonance field will sooner or later manifest physically. To put this in simple terms we can say that the whole solar and stellar system forms a totality of thousands of resonance maps which facilitate permanent dialog between the earth and the other members of this system. Ultimately all this is just information which will eventually take form in matter in some way.

Forces active behind the earth's physical composition

At the physical level the earth is composed of minerals, plants, animals and human beings, all of which are accessible to our senses. But let us not forget that these physical manifestations are composed of particles that are assembled and densified by specific informational forces.

What is it that gathers two hydrogen atoms with one oxygen atom so as to produce water? This is the important question. Water is not just composed of two molecules of hydrogen and one of oxygen as we are taught (so that we can become an agricultural engineer or a vet kowtowing before the might of the chemicals industry). *Besides its constituent atoms water is also the force which gathers them together.* Without this there would be no water! We can demonstrate this in a more down-to-earth example: would you prefer to eat a cake or its separate constituents? A chef or cook is needed to gather the ingredients in a creative act, to produce a successful dish. One would not dismiss the work of the cook, but praise him as the originator of the cake. So why should we hide acknowledgement of the active or formative forces as they are known in biodynamics? Are we worried that these discoveries will undermine the whole gigantic market in artificials? What do these forces do which are used by biodynamics, homeopathy, music therapy and many other methods that go back to primordial times, whose beneficial effects are becoming ever more widely recognized? They connect matter with the resonance of its originating formative plane.

The two solar aspects

So how do we use these formative forces? This is also something that Rudolf Steiner dealt with in his Agricultural Course, and in many other works which relate to the sources of life. Life on earth is nothing but a dialog between visible and an invisible world; an invisible, formative world and a world that is formed and is accessible to our physical senses. This interaction occurs through continuous 'incarnation' or physical embodiment of centripetal forces, and ascent or disembodiment of centrifugal forces. When a plant, after forming its stem and blossom, arrives at the pollen-forming stage, it has almost attained a stage of non-matter largely emancipated from gravity. This allows pollen to fly up into the air to a height of several thousand meters – something that a root wholly subject to centripetal forces cannot do. Annuals thus complete their cycle by being partly dematerialized, firstly, and then contracted to an extreme in the seed. Thus in autumn one sees separate the forces which spring had united. All around us centrifugal forces ceaselessly interpenetrate with centripetal ones. Aromas, tastes and colors can only descend to the physical plane when forces are reversed, that is to say when a process of growth has come to a halt. Without this, fruit cannot form properly, something aided and supported by one of the biodynamic preparations. Of course these two stages are not totally separate: at a certain moment they interpenetrate. But we cannot achieve a fine taste when growth has been over-forced, something which ultimately opposes the process of latent contraction that always underlies fruiting.

It is for this reason that the vine needs to be in something of a combat situation, so that its growth is reined in to produce a good wine. The fruiting forces descend more easily where growth is weaker. In fact, one of the biodynamic preparations contributes to this. Thus we cannot properly understand how life manifests on earth without always trying to take account of these two opposing forces. In the same way each planet has a current which approaches matter and aids the plant's physical incarnation; and another, opposite current which first halts its growth and then encourages the process of dematerialization. The action of the sun on earth is also sensitive to these two aspects of life which one can call spring like and autumnal, even though they do not always adhere to the seasonal cal-endar. One can even say that it is the sun that activates these. Let us never forget, though, that the sun is less an agent than a coordinator.

Let us recall, also, that two biodynamic preparations sup-port these two solar aspects. On the one hand the horn dung which mobilizes matter in the plant, and on the other the horn silica which forms and sculpts this matter. Every other biodynamic preparation – based on chamomile, yarrow, net-tle, oak bark, dandelion and valerian – acts as a connection to the archetypal forces of the five planets and the moon. Biodynamics is just an enhancement of the solar system's manifestations on the earth, at a time when, due to our grave lack of understanding, its influences have been drastically weakened by the energy dustbin we are making of our atmosphere. Above all, as Steiner impresses on us, the atmosphere is the place of interaction between earthly and cosmic laws, without which the earth would be a corpse. The atmosphere is very organized in successive strata. Energy pollution of the atmosphere (physical pollution is actually less grave) weakens the life forces which sustain the earth and thus many of those which also sustain the human being. This is why biodynamics has become so important: it responds at the energy level to an alarming imbalance. It reconnects the earth to its sources. This is also how we can understand its effect on radioactivity, to which it represents an opposite pole. It has an incarnating effect whereas radioactivity aids the escape of life. Natural radioactivity is involved in a process of ageing that is necessary to evolution and even beneficial to the human being in certain circumstances: consider the number of artists in Prague, where natural background radiation is very strong. When man produces it, however – or rather when we liberate the electrons that imprison heat dandified by earthly forces – it becomes devastating.

In Poland, at the time of Chernobyl, on a biodynamic commune, scientists measured only 1/10th of the radioactivity which was present elsewhere. Two hundred years ago biodynamics would probably have had less relevance. Today it has become essential, firstly in relation to taste, to wine and food quality, but also in relation to the life which the earth needs to maintain its equilibrium. The climate, after all, is composed of nothing but balances and harmonies. The troubles which plague it cannot be ascribed to warming alone. It is the energy system organizing the climate which has been so unsettled and disrupted.

An information system

The solar and stellar system emits forces which form and sculpt matter, or which weave its threads together. According to the season it has a dandifying or dissipating effect. In delivering energy from the prison of matter, we open a door to the pure forces of life. The more dilute a substance, and the less matter it contains, the greater are the energy effects. Biodynamics draws on and reinforces the archetypal forces of the solar system which a plant needs. These two opposing but also complementary levels, of the tangible and intangible, are jointly active all around us. *Both are necessary for our understanding of life on earth. The one should not conceal the other.*

Thus we should regard the solar system as an 'information' system whose task is to generate on earth organisms which conserve the forces of life while also coming to physical manifestation. When these life forces are extinguished by matter's density they can no longer survive as they no longer penetrate matter sufficiently. This is what we call death. This just means a departure of life from the physical plane. The energy plane cannot die, but only transform. This is what Goethe meant when he said that nature invented death to recreate life:

As long as you have not grasped
Life issuing forth from death
You're but a troubled guest
Upon the shrouded earth.

The solar system is composed, as we know, of the five planets, the moon our satellite, and the sun which acts rather like the conductor of an orchestra. Each planet plays a very precise role in relation to plants, their growth and maturation. Like the earth, each one both emits and receives. Scientists know that Jupiter, for example, is a vast radio-electric receiver sometimes more powerful than the sun. It can emit more than it receives when one of its satellites occupies a particular position relative to the earth, the sun and Jupiter itself. We can understand this in terms of the way we turn an antenna to get better reception. The movement of the planets around the earth is very similar. In other words there are stronger and weaker 'reception' positions. In relation to the phase of growth of each plant, the work it accomplishes at a particular moment, and the beneficial or inhibiting positions of the planets relative to the earth, either beneficial or injurious influences will affect the quality of wines or vegetables. The 50,000 morphochromatography images made by André Faussurier when he was lab director at the Catholic University of Lyon revealed a correlation between astronomical disturbances (eclipse, lunar node, occultation of one planet by another etc.) and irregularities of the images he obtained. Lakovsky, mentioned above, who worked at a period when technology had not yet started to modify the taste of wine, demonstrated clear correlations between years when sun-spot activity was at a maximum and great wine years. We also know that aurora borealis can have a very beneficial effect on grain harvests, something shown by Maria Thun in Germany in her books and invaluable calendars. What is grudgingly conceded in the case of the moon is also true for the whole solar system. The moon's reflective action is somewhat like an intermediary between the earth and the solar system. It exerts an influence on all electromagnetism, and thus on all that affects life on earth. For example, breeding either at full or new moon helps determine the sex of an animal. Having a female covered by a male on the day of either the full or new moon gives an 80 percent likelihood that, respectively, a male or female offspring will be born. Biodynamic breeders actually already know something about this. From new moon towards full moon, moon forces draw plants upwards. From full moon to new moon, in contrast, forces are progressively orientated earthwards, and thus stimulate root growth.

A living system

We can now present the solar system as a living organism and not as an engine in which all the parts just keep repeating the same movements continuously, for in life nothing repeats exactly in the same way. The revolutions of the planets, their years if you like, always vary a little. That of Mercury, the most irregular, can vary by eight days. The sun itself is never exactly at the center of the more or less elliptical orbits that the planets describe around it. Their rotational planes in relation to the sun can also reveal differences, as can their speed of orbit. All this really testifies to a living organism. This collective dance of the earth and the planets around the sun – a little like the electrons compelled to encircle an atom – is full of irregularities and thus, as we will see, often provides a wealth of opportunities for viticulturists or farmers who know how to perceive and use it. All these irregularities also help us to understand that each day has a unique, unrepeatable aspect, which hopefully renders us more attentive to its particular characteristics.

Using this knowledge

There are two simple, basic rules for making use of this knowledge. These draw on the four states of matter which we mentioned at the beginning. Firstly we can say that the planetary or stellar impulses express themselves in four principle ways, linked to these four states of matter: that of fruiting (heat impulse), blossoming (light impulse), leaf growth (water impulse) and finally root growth (earth impulse). The second rule is that, according to the positions of the planets in relation to the earth, and in relation to the stellar constellations present behind them as they orbit the sun, these different influences act in a more or less pronounced and harmonious way. In specific instances, such as an eclipse, lunar or planetary nodes (when planets cross the sun's plane), lunar apogee or perigee positions (when the moon is furthest from or nearest to the earth), these can have some very adverse effects.

Of course the whole earth is always subject to these influences, whether one studies them in chemistry, biology or biodynamics. But when agriculture has killed the majority of living agents such as a soil's micro-organisms, or has modified the growth of a plant through artificial interventions such as chemical fertilizers or dangerous systemic herbicides etc., the plant becomes less receptive, or not receptive at all, to these celestial influences. It is like someone falling ill and losing some of his capacity for communication. In consequence the plant becomes somewhat deaf to these invisible, qualitative planes of life, and this is why the resonance qualities of so much of our food are so poor for human nutrition – as quality tests, called 'sensitive crystallization,' have shown. These tests are done with copper chloride, as powder, mixed with the juice of any food, and enable us to observe a product's energy organization. The liquid is placed on a piece of glass, and then in a controlled-humidity Petri box. As it dries up the product's life forces draw the powder in lines, producing a shape somewhat similar to frost flowers on a window. If the product is devoid of life forces no image will arise but merely a spot of dried-up powder. If it is alive, the resulting image can be analyzed and the quality of the product determined by the pattern's regularity, peripheral structure and the precision at its center (see Plate 13).

The tests are also used on human blood, to ascertain the energy effects on it of any medicine prescribed. The images produced embody life forces. As we have already seen, what feeds us in our food are these forces rather than mere matter, and impoverished food will show little such life. In other words, an unsound agriculture isolates the plant from the vast context which gives it life, and, as we have already discussed, leads to the need to give it continuous artificial assistance. All living beings on earth, starting from the human being and his powers of cognition, but also all the plants, animals, insects, birds and micro-organisms of all kinds, even if they are minuscule, have a role to play in linking the earth to the macrocosm for which they act as intermediaries.

These insights can help each one of us to question conventional agricultural practices. When we speak of the macrocosm and its constituent aspects, we cannot do so without acknowledging the 40 years' work undertaken in Germany by Maria Thun, with such rigor and courage. She has painstakingly measured each planetary and stellar influence on the plant world, and now regularly publishes a sowing and planting calendar as a guide for others.[22]

This agricultural vade mecum has opened up a whole field for further research. Isn't it interesting to observe, for instance, that several generations of wheat grains tuned fully to their proper planetary and stellar influences by a conscientious biodynamic practice, reduce wheat-related illnesses to almost nil. Thus wheat has recovered its former nutritious properties, and can once again nourish us fully. Such observations also apply to other plants. And all this knowledge is making consumers more aware of the economic interests underpinning the artificial interventions employed by conventional agriculture.

If we take account of the positions of the planets in relation to each other, and to the stellar constellations, in this narrow band which includes the rotation of all the planets around the sun – which we call the zodiac in fact; and of the continuous, very rapid movement of the sun and our solar system towards the constellation of Vega; of the behavior of the sun itself with its sudden eruptions; and of many other things as well, then we can see that each day really is unique and

---

[22] Maria Thun, *Sowing and Planting Calendar* (Floris Books); See also *Gardening For Life* (Hawthorn Press, 1999) and *The Biodynamic Year* (Temple Lodge 2000).

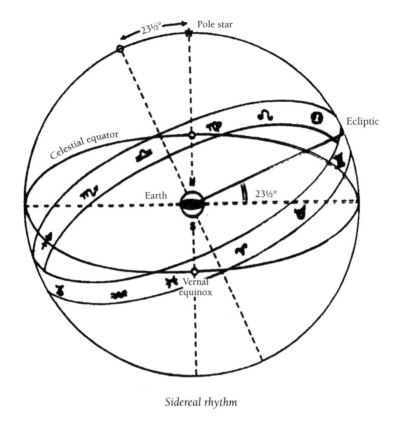

Sidereal rhythm

unrepeatable. This means in turn that all the influences which reach us from the solar system via the living medium whose rediscovery began in 1831 with Faraday's radio waves – which he named 'the medium which links the emitting and the emitted principle' – are in permanent flux. Every angle of planets between each other and in relation to the earth has an energy effect differing according to numerical laws. (Forces linked, for instance, to the number 2, or 3, or 4 or 5, do not exert the same effects.) These numerical laws are the same as those that form the Platonic Solids. The same principle is, ultimately, at work in the construction of great architectural creations, but here the position, relationships between angles, patterns and proportions are not fixed but change from one minute to the next. Thus we need to try to accentuate our 'hearing' for the energy utterance that speaks the same language as the vine, and that reinforces and harmonizes the particular work it is accomplishing at any particular time.

In using these facts we need to try to orientate each receiver (the different plants) to the planetary and stellar situation that is best for each one. Thus there are favorable and unfavorable moments. There are kinds of 'information' which tend towards the harmonious formation of matter, and other kinds which oppose it. When we seek to generate a greater degree of life in a particular area we need to remember this. The ascending or descending cycle of a planet does not generate the same conditions. We know this from observing the effects of the moon, but we also need to understand it at the level of all the planets, and discover health-giving forces there for every species of tree, each of which is always subject to a particular prevailing planetary influence. One sows a tree when its planet is in the ascendant, and transplants it when the planet is in the descendant. The principle is the same as the effect of the ascending moon on the plant's vegetative parts, and of the descending moon on a plant's roots. By transplanting at the right moment (which requires us to know the planet that governs a particular tree) we can tune a tree to its specific archetypal forces, helping the roots to recover from the shock of transplanting and giving it much greater health and resistance to disease.[23] For distant planets like Saturn, it is sometimes worth waiting 15 years for the best moment to plant. I do not intend to give all the details of these procedures (see Bibliography for further reading) but to convey a vivid understanding of their rich diversity. The mercurial vine is best transplanted when Mercury is in the descendant.

All the particularities of the system in which we are embedded thus endow each day with a unique energy information profile, enabling us to ask how we can enhance certain effects, if we consider them desirable, in relation to our vines. All this knowledge produces results which each person can try out for himself and measure. There are innumerable examples. Where the laws which generate life can fully unfold, each plant reacts to the least influence with which one connects it: rather like pupils in a class who respond to the all the hints and suggestions of a teacher who knows how to present his knowledge in a way they can relate to. We are surrounded by examples which encourage us to take this kind of approach. It has already been scientifically demonstrated that certain people can speed up or slow down the growth of a plant through their very presence and way of thinking. Such heart forces should

---

[23] Maria Thun demonstrated this clearly several years ago in her study of pine plantations. Sixty years after planting, some trees were in a very poor condition while others close beside them were unaffected by diseases.

definitely not be overlooked. Crystallization techniques have shown that plants modify their behavior 24 hours before a solar eclipse. We know that some animals sense the approach of an earthquake more than a day before it happens. We also know the reality of the so-called 'placebo' effect on human beings – which ultimately is information we unconsciously give ourselves. This new path of research is currently being pursued further by certain scientists who not only study this system of energies but are also starting to use its potential effects. These discoveries can lead to huge progress or problems, depending on the conscience of those who apply them. And here lies the nub of the problem. Are we ready for this kind of knowledge? Here's one example: intensive pig farming creates enormous problems of smell and pollution for several kilometers around. By giving each animal 2 grams of quartz per day that has been treated with certain alpine plants, the dung becomes normal again, and the animals' hides regain their glossiness. But do we have the right to make pigs believe they are in a mountain environment when in fact they are reared in an animal concentration camp? In other words, knowledge of how to use energy forces could lead to the most absurd practices. Instead of striving for an animal's wellbeing by increasingly helping it to live in harmony with its archetypal energies, such knowledge could just be used, without conscience, for profitable ends. Similar tricks could be used in viticulture, but are quite at odds with an overall context of care for the natural world. For instance, it is possible to imprint water with energy information: adding 2 or 3 drops of such water will immediately improve the quality and taste of wine. But this is a kind of deception, for it is a later intervention rather than an integral part of the whole process. The effect will not be lasting and will be reversed at a later stage. So you can see that working with energies in this way could, alas, have an important economic potential. Instead, biodynamics aims to reinforce a connection that exists already rather than creating one arbitrarily.

Practicalities

Let us return to specific practice, though, and ask what actions are necessary to enhance planetary and stellar influences on a vineyard.

Quite simply, when a situation is propitious for the vine we will try to reinforce its influence. What do I mean by a favorable situation? There are three zodiac constellations which give heat impulses, and thus fruiting forces which the vine uses to form the grape. These constellations are: Ram, Lion and Archer.[24] We will not enter here into the specific quality of each of these influences. Those interested in these matters can read the excellent book by Fritz Julius referred to in the Bibliography, or the numerous books by Maria Thun; or, equally, astute, the book by Kranich. But to cut a long story short it

---

[24] The English names are used, rather than the Latin, to highlight the fact that the astronomical reality is referred to, rather than astrological tradition (see further comment in text below).

is good, whenever possible, to enhance the vine's connection with these three influences, so as to develop the wine's full complexity. This is called 'trigon treatment.' Of course everyone is free to disagree with such an approach. Let us remember, though, that we are using astronomy, not astrology here – i.e. the actual position of the constellations, which differs by almost 30 degrees from that of astrology. We have to go back over 2,000 years to find a situation where astrological positions coincided with reality (although this does not mean that the latter no longer have any effect on the human being today, as we have passed through a long evolution whose memory is stored in our bodies. There is continual interaction between the past, the brief, beautiful present and the future).

Apart from these constellations there are two other heat or warmth – and thus fruiting – influences, this time via the planets Mercury and Saturn. The first has a shorter year than ours, and the other a year that is much longer (30 years). Their effects are likewise very different. Mercury affects movement, sap circulation, growth, physical accumulation, and helps avert hindrances or illnesses. Saturn, by contrast, plays an important role in taste, maturity and concentration. The quality one admires in a wine is derived from Jupiter or Saturn. The book by Kranich (see Bibliography) offers many details. We would not have juicy and tasty fruits without these two influences. Jupiter helps juice formation in a fruit, while Saturn creates a concentration of taste. The key question for the viticulturist is how and at what moment to enhance them. It requires fine sensitivity to decide the right moment to harvest the grapes. By waiting for the grapes to start shrinking and concentrating you lose a great deal of juice (decreasing yield) but the taste shifts from fruity to a kind of mineral taste. At the same time the color changes from a light, yellowish-green to a deep yellow, sometimes almost orange, giving rise to the depth and essential quality of an AOC.

Let's take an example: when a warmth planet, either Mercury or Saturn, is situated in front of a heat constellation, their combined effects are enhanced and enriched by this synergy. And apart from this, when the position of one or both of these planets in relation to the earth is, for example, 120° or 180° – the latter being called an opposition – the earth which is at the center of this straight line linking the two planets receives their influences in an enhanced way.

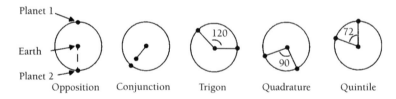

Different angles of positive (180°, 60°) or negative conjunctions (quadrature, quintile) of two planets with the earth.[25]

---

[25] From: Maria Thun's *Sowing and Planting Calendar,* Floris Books

There are other beneficial situations, but this must suffice as one instance of many. By tuning into and taking advantage of such effects, a biodynamic vineyard will benefit from them. What is also interesting here is that we can *enhance our vines' capacity to receive these influences,* for example by working the soil or dynamisation. Even light hoeing of the ground – that is, opening the skin of the soil – on a favorable date will render the micro-organisms more sensitive to these cosmic influences. It is like opening the shutters in a room to get a better idea of the weather outside. We can also choose such a date for dynamising a preparation. The vortex created in this process will thus imbue the substance to a greater extent with the specific, prevailing situation on that day, and then this information will be passed to the leaves or the soil. The vine, which has something of a nerve's sensitivity, is very receptive to these effects. But again, this is only one aspect of biodynamics, which is primarily effective through systematic application of its preparations once or twice each year. We can therefore practice it without any knowledge of the solar system. However the choice of date, also of course taking weather conditions into account, is an important complementary aspect. All these things have specific, measurable effects on yield, resistance to disease, storage quality, taste and aroma, and lead to very significant differences. Ultimately we are just helping the plant to express itself better by connecting with underlying planes of energy or archetypal forces. This is parallel, in the field of music, to improving the acoustics of a room, or the sound qualities of an instrument.

When we try to understand the profoundly original nature of each planet and each constellation, we acquire the possibility of creatively affecting the behavior of our vines. Thus we can help them imbue themselves better with one or another influence which seems to us the right one for our vineyard. For example, we can understand that, when the sun is in the constellation of Archer in the depths of winter – thus at a time when nothing much emerges from the earth – the prevailing centripetal effects are very different from those of the Ram, which carries the strong centrifugal impulses of spring.

Let us also remember the moon which, in orbiting the earth, passes through the zodiac in 28 days, enabling us to use each of these specific forces during at least part of each month.

Plate 16 shows the importance of dates for harvesting and planting. The top picture is of identical vines, cut and planted in the same soil at an interval of several days. Some grow strongly while others fail to root. The bottom picture shows lettuces planted on either light (blossom) or water (leaf) days, at an interval of several days. The blossom day produces lettuces that tend to 'shoot' into flower. (Source: Maria Thun: *Sowing and Planting Calendar*)

This approach represents a return to a more artistic understanding of agriculture. Each day we can affect plant growth by means of our human qualities, and through simple, small actions based on our knowledge of the solar and stellar systems and the life they engender.

All this, as we have seen, is enhanced by the physical nature of a landscape, the choice of appropriate and well-raised vine varieties, specific animals adapted to the location, and many other factors which all act as receiver antennae for these impulses of life. We can regard all these factors as the musical notes which form a complete work. The more harmonious chords there are the more vital and vibrant our wines will be. Resistance to oxidation in biodynamic wines, that is, to destructive forces of ageing – which should not be confused with full maturity – is already a real demonstration of these facts. It is here that hasty tastings do not allow true wines to unfold their potential. When things are done in a rush, the 'flashy' and 'cosmeticized' wine will grab your attention.

This quick summary of biodynamic viticulture has the sole aim of showing wine lovers that it is not some superstitious sorcery based on an illusory view of the world, nor a wily communication ploy, but rather a reality which each of us comes to in our own way, in our own good time, to the extent we penetrate it with our understanding. Over time this agricultural approach, involving a greater discovery of the laws of life rather than those of matter alone, will increasingly gain ground. This is really the only way of offering consumers a healthy product, whose distinctive tastes are imbued with the landscape where it grows. Such produce is free from the artificial palate-flattery which we all too often think derives from the earth. If such artificial taste enhancement remains legally acceptable, at the risk of destroying the great concept of the AOC, at least this should be noted on the label so as to give consumers an opportunity to know whether what they are buying is an authentic taste of *terroir*. Isn't that the least of their rights?

6

Conclusion

When you drink a real wine, when you are transported by particular tastes or aromas, it is really a far-off, ethereal world that you are admiring, one distant from earthly laws. Each biodynamic agricultural act respects and sustains this other reality, transforming it into a physical quality which thereby becomes perceptible to our senses. By extending our knowledge, by giving back to the earth all its faculties through a respectful and artistic agriculture, the human being can come to play his full role. Then we will be able to receive, in return, more creative impulses from this world of resonance, whose nourishment we draw on unconsciously every day through all our senses, including sight. Aesthetic things are also nourishment, something to which we should give more consideration in education. Children are too often exposed to ugliness of all kinds, which too easily becomes the norm for them. A person well-nourished by a world of resonance fully expressed and materialized will be in greater harmony with the forces that gave birth to and sustain him. The viticulturist would say: 'You are what you drink,' and the farmer would say: 'You are what you eat.'

This is a way to understand the legend of Dionysus, torn to pieces by the Titans on the orders of Hera, thus incarnating his forces of human individuality and compelling our descent to earth, where individuality can unfold. The Titans break the energy totality by ripping apart the body of Dionysus, thus making human beings enter time, the forces of Chronos, and leave Uranus, his father, the symbol of a unified whole.

Or, if you prefer, we have left the macrocosm and submerged ourselves in the fragmented microcosm. This separation created some dramas of course, and many partial and therefore incomplete and dangerous discoveries. In thinking ourselves masters of the world, we have brought this world close to the edge of destruction. But the legend of Dionysus is full of optimism. It tells us that his heart is saved by the goddess Pallas Athena, and entrusted to Zeus. A love affair brews which leads to the union of a 'mortal,' Semele, with Zeus. From these new forces Dionysus is reborn, and subsequently takes on the role of teacher of agriculture, and the science and cultivation of the vine. This image shows us the rebirth of the human being through his heart forces, and through accessing a macrocosmic knowledge which we can put to good use. Biodynamics is Steiner's contribution to this rebirth, of which the younger generation is in such urgent and desperate need if it is to find its way back to a meaningful, sane and health-giving relationship with both earthly and ethereal realities.

Epilogue: A perspective from an admiring chemist-winemaker

Several years ago, I visited Nicolas Joly's Biodynamic vineyard in the Loire Valley and tasted his great wines. I was very impressed both by their quality and their unique expression.

Although Biodynamic theory does not exactly "cope" with the rigors of the scientific method, its practice in grape growing certainly shows unique results. Some of its key ideas about natural phenomena are expressed in classical symbols, which the ancients used to help describe humanities place in the universe. In other words, Biodynamic notions "humanize" nature. In his book, Joly relates to the natural vine phenomena like bud burst, veraison, grape maturation, and the whole annual cycle, as if it were the expression of an independent personality. This angle of observation is different from those practiced in the natural sciences, namely, every statement has to be proven, either by experiments or by careful observation if experiments can not be done. Joly frames his concepts as basic truth, without the benefit of empirical proof. The methods and treating preparations used to grow grapes by Biodynamic guidelines may seem to fall in the zone between fiction and science. But so does homeopathic therapy, and it works well in many cases.

I fully accept most of the practical concepts in the book, which can be summarized as minimal intervention in the wine making process. I accept them because they are, at their core, ethical principals.

By this I mean that wines should be unique and authentic. They should carry their terroir and personal characteristics with as minimal human manipulation as possible.

In this, Joly provides very important messages. He criticises the use of chemical fertilizers, herbicides, fungicides, pesticides, and other questionable vineyard techniques. He explains that cellar practices such as reverse osmosis, micro-oxygenation, cryo-extraction, and the use of isolated yeast, oak chips, enzymes and additives, mask the unique expression of terroir. What are the results of all of this intermediacy? Standardization.

Two years ago I attended a tasting carried on by a very well know international consultant. We tasted 15 wines from very different locations around the globe. All of them were wonderful. But it didn't take long to realize that they all tasted as if they originated from the same place. Joly sums up this all-too-common scenario quite nicely: "Here we find good, false wines, for which the AOC is really just writing on the label rather than a reality in the bottle. The full, original taste, which each 'Appellation controlle' once guaranteed to the consumer, no longer exists."

While reading this book, I recommend being open-minded to this different system. Some of the theoretical concepts are difficult to comprehend. You may decide to accept Biodynamic theory in full or in part, those that ring true to your own experience as grape grower or winemaker or both. For me, it was eye-opening just how much there is in Biodynamics for a chemist winemaker, or should I say, winemaker chemist, to identify with.

Yair Margalit

Caesarea, Israel

January 12, 2008

Yair Margalit, Ph.D. is a physical chemist with more than 30 years of research and teaching experience. He operates one of Israel's premier wineries, which bears his name, and teaches winemaking and wine chemistry at the Israel Institute of Technology—the "Technicon." His books include *Winery Technology and Operations, Concepts in Wine Technology* and *Concepts in Wine Chemistry*

Appendix

I. The 'Return to Terroir' Association

This group, created in France in 2001 by Nicolas Joly, now comprises 153 winegrowers from 13 different countries. Its purpose is to guarantee the full validity and expression of the 'appellations,' and ensure wines of high quality and great originality.

To achieve this we act at three levels:

1. A legal guarantee of good agricultural practices. This means organic and/or biodynamic certification for the whole vineyard, for at least the past three years. The wine bearing this label comes from a vital soil that has not been treated with chemicals. The consumer thus has a legal assurance of quality. Ninety percent of group members practice biodynamic agriculture.

2. A guarantee that no actions undertaken in the cellar change the full expression of the AOC's taste. Proper agricultural practice means the cellar can be a a birth or fruition procedure rather than a factory. All 300 aromatic yeast are banned, as is osmosis, GM, mechanical harvest (see quality charter). The winegrower signs a commitment to cellar procedures covering the past 3 years.

3. Wines are tasted by a committee consisting of the following well-known winegrowers:

Alsace: Olivier Umbrecht

Bourgogne: Anne Claude Leflaive and Pierre Morey

Bordeaux: Jean Luc Hubert

Champagne: David Leclapart

Rhone: Philippe de Blicquy

South: Raymond de Villeneuve

Loire: Nicolas Joly.

New members can only join the group by the committee's unanimous decision.

Three to four tastings are organized worldwide every year, and the costs of this are shared by all group members.

You can find more information at our website (in construction): www.biodynamy.com

II. The Charter of Quality

The system of evaluation outlined below does not speak in terms of 'biodynamic' or 'non-biodynamic,' but simply of actions which permit an appellation to express itself fully. Thus one can go from one to three green stars, adding to this the usual notations used by wine guides.

This system encourages winegrowers to aim for the highest standards, and informs our expression.

One star:
Wine from a controlled appellation of origin has a particular taste linked to the type of soil and climate. Agriculture should therefore enhance the organic life of the soil and avoid all synthetic chemical products.

* No weed-killers/herbicides
* No chemical fertilizers
* No synthetic chemical products
* No systemic treatments
* No aromatic yeasts

Two stars:
In recent years the formidable advance of technology has enabled winegrowers to re-create the tastes that an inadequate agriculture had drained from grapes. A return to good practices renders this technology pointless, restoring the original taste to each wine without misleading the consumer.

* No mechanical harvesting
* No extraneous yeast that is foreign to the location
* No treating of the must with enzymes
* No concentrator that works by inverted osmosis
* No cryo-extraction
* No cold treatment that reaches freezing point

Three stars:
* No de-acidifying or re-acidifying
* No addition of ascorbic acid, nor of potassium sorbate
* No chaptalization, including concentrated must
* No irrigation

All winegrowers who adopt this charter will have authentic and thus inimitable wine since the relationship between soil and climate has a different 'countenance' in each location. Viticulturists who sign the charter in the presence of a notary public commit themselves to respecting this code of ethics in relation to the entire production cycle. Inspections can be carried out at any time by a member of the group.

Return to Terroir, Members 2007

France

Alsace

Jean Pierre et Chantal FRICK – Domaine Pierre Frick – 5 rue de Baer – 68250 Pfaffenheim
Tel: 03 89 49 62 99
Fax: 03 89 49 73 78
E-mail: pierre.frick@wanadoo.fr (en biodynamie)

***

Jean Miche DEISS – Domaine Marcel Deiss – 15 route du vin – 68750 Bergheim
Tel: 03 89 73 63 37
Fax: 03 89 73 32 67
E-mail: marceldeiss@marcel deiss.com (en biodynamie)

***

Léonard et Olivier HUMBRE-CHT – Domaine Zind Humbrecht – 4 route de Colmar – 68230 Turckheim
Tel: 03 89 27 02 05
Fax: 03 89 27 22 58
E-mail: o.humbrecht@wana doo.fr (en biodynamie)

***

Jean et Béa SCHAETZEL – Domaine Martin Schaetzel – 68770 Ammerschwihr
Tel: 03 89 47 11 39
Fax: 03 89 78 29 77
E-mail: jean.schaetzel@wana doo.fr (en biodynamie)

***

Marc TEMPÉ – 16 rue du Schlossberg – 68340 Zellenberg
Tel: 03 89 41 20 38
Fax: 03 89 23 21 60
E-mail: contact@lasommeliere.fr (en biodyanmie)

***

André OSTERTAG – Domaine Ostertag – 87 rue Finkwiller – 67680 Epfig
Tel: 03 88 85 51 34
Fax: 03 88 85 58 95

E-mail: domaine.ostertag@wanadoo.fr (en biodynamie)

***

Marc KREYDENWEISS – Domaine Kreydenweiss – 12 rue Deharbe – 67140 Andlau
Tel: 03 88 08 95 83
Fax: 03 88 08 41 16
E-mail: marc@kreydenweiss.com (en biodynamie)

***

Jean MEYER – Domaine Josmeyer – 76 rue Clemenceau – 68920 Wintzenheim
Tel: 03 89 27 91 90
Fax: 03 89 27 9199
E-mail: christophe.ehrhart@wanadoo.fr (en biodynamie)

***

Marie ZUSSLIN – Domaine Valentin Zusslin – 57 Grand'Rue – 68500 Orschwihr
Tel: 03 89 76 82 84
Fax: 03 89 76 64 36
E-mail: domaine.zusslin@wanadoo.fr (en biodynamie)

Bourgogne

Marc LAFARGE – Domaine Michel Lafarge – Rue de la Combe – 21290 Volnay
Tel: 03 80 21 61 61
Fax: 03 80 21 67 83 (en biodynamie)

***

Aubert de VILLAINE et Pierre de BENOIST – A et P de Villaine – 71150 Bouzeron
Tel: 03 85 91 20 50
Fax: 03 85 87 04 10
E-mail: dom.devillaine@wanadoo.fr (en biologie)

***

Lalou BIZE LEROY – Domaine Leroy/Domaine D'Auvenay – 15 rue de la Fontaine

21700 Vosne Romanée
Tel: 03 80 21 21 10
Fax: 03 80 21 63 81
E-mail: domaine.leroy@wanadoo.fr (en biodynamie)

***

Emmanuel GIBOULOT – Domaine Giboulot – 4 rue de Seurre – 21200 Beaune
Tel: 03 80 22 90 07
Fax: 03 80 22 89 53
E-mail: emmanuel.giboulot@wanadoo.fr (en biodynamie)

***

Anne Claude LEFLAIVE – Domaine Leflaive – Place des Marroniers – 21290 Puligny Montrachet
Tel: 03 80 21 30 13
Fax: 03 80 21 39 57
E-mail: sce.domaine-leflaive@wanadoo.fr (en biodynamie)

***

Pierre MOREY – Domaine Pierre Morey – 13 rue Pierre Mouchoux – 21190 Meurault
Tel: 03 80 21 21 03
Fax: 03 80 21 66 38
E-mail: morey-blanc@wana doo.fr (en biodynamie)

***

Jean Louis TRAPET – Domaine Trapet Pere et Fils – 53 route de Beaune – 21220 Gevrey Chambertin
Tel: 03 80 34 30 40
Fax: 03 80 51 86 34
E-mail: message@domaine-trapet.com (en biodynamie)

***

Benjamin LEROUX – Comte Armand, Clos des Epeneaux – 7 rue de la Mairie – 21630 Pommard
Tel: 03 80 24 70 50
Fax: 03 80 22 72 37
E-mail: contact@domaine-des-epeneaux.com (en biodynamie)

***

Didier MONTCHOVET – Domaine Montchovet – Rue de l'Ancienne Gare – 21190 Nantoux
Tel: 03 80 26 03 13
Fax: 03 80 26 05 19
E-mail: didier.montchovet@wanadoo.fr (en biodynamie)
***

Julien GUILLOT – Domaine des Vignes du maynes – 71260 Cruzille
Tel: 03 85 33 20 15
E-mail:info@vignes-du-maynes.com (en biodynamie)
***

Dominique DERAIN – EARL Catherine et Dominique Derain – 46 rue des Perrières – 21190 St Aubin
Tel: 03 80 21 35 49
Fax: 03 80 21 94 31
E-mail: dc.derain@wanadoo.fr (en biodynamie)

Bordeaux

Alan Geddes – Château Mayrague
***

Alain Moueix – Château Fonroque – 33330 Saint Emilion
Tel: 05 57 24 60 02
Fax: 05 57 24 74 59
E-mail: mazeyres@wanadoo.fr (en biodynamie)
***

Pierre Blois – Moulin du Cadet – SAS Blois Moueix – 33330 Saint Emilion
Tel: 05 57 55 00 50
Fax: 05 57 51 63 44
E-mail: moulinducadet@wanadoo.fr (en biodynamie)
***

Alain FERRAN – Domaine Ferran – Le Tucaou – 33760 St Pierre de bat
Tel: 05 56 61 98 61 / 06 75 20 21 02
Fax: 05 56 61 97 84
E-mail: alain.ferran@tiscali.fr (en biologie)
***

Johnet Véronique COCHRAN – Château Falfas – 33710 Bayon sur Gironde
Tel: 05 57 64 80 41
Fax: 05 57 64 93 24
E-mail: jvcochran@online.fr (en biodynamie)
***

Claire LAVAL – Château Gombaude Guillot – 4 chemin des Grandes Vignes – 33500 Pomerol
Tel: 05 57 51 17 40
Fax: 05 57 51 16 89
E-mail: château.gombaudeguillot@wanadoo.fr (en biologie)
***

Louis LURTON – Château Haut Nouchet – 33650 Martillac
Tel: 05 56 72 69 74
Fax: 05 56 72 56 11
E-mail: chateau-haut-nouchet@wanadoo.fr (en biologie)
***

Olympe et yvon MINVIELLE – Château Lagarette – 33360 Camblanes
Tel: 05 56 20 08 78
E-mail: château.lagarette@wanadoo.fr (en biodynamie)
***

Paul et Pascale BARRE – Château La Grave – 33126 Fronsac
Tel: 05 57 51 31 11
Fax: 05 57 25 08 61
E-mail: p.p.barre@wanadoo.fr (en biodynamie)
***

Jean Pierre AMOREAU – Château Le Puy – 33570 St Cibard
Tel: 05 57 40 61 82
Fax: 05 57 40 67 65
E-mail: amoreau@chateau-le-puy.com (en biologie)
***

Jean Luc HUBERT – Château La Grolet – 33710 St Ciers de

Canesse
Tel: 05 57 42 11 95
Fax: 05 57 42 38 15
E-mail: peybonhomme@terre-net.fr (en biodynamie)
***

Alain DEJEAN – Domaine Rousset Peyraguey – 8 Lieu dit Arrançon 33210 Preignac
Tel: 05 56 63 49 43
Fax: 05 57 31 0833
E-mail: ROUSSET.PEYRAGUET@wanadoo.fr (en biodynamie)

Champagne

Jean Pierre FLEURY – Champagne Fleury – 43, grande rue – 10250 Courteron
Tel: 03 25 38 20 28
Fax: 03 25 38 24 65
E-mail: champagne-fleury@wanadoo.fr (en biodynamie)
***

David et Carole LÉCLAPART – Champagne David Léclapart – 10, rue de la Mairie – 51380 Trépail
Tel/Fax: 03 26 57 07 01
E-mail: david.leclapart@wanadoo.fr (en biodynamie)
***

Françoise BEDEL – Champagne BEDEL – 71 Grand Rue – 02310 Crouttes sur Marne
Tel: 03 23 82 15 80
Fax: 03 23 82 11 49
E-mail: chfbedel@champagne-francoise-bedel.fr (en biodynamie)

Corse

Pierre Richaume – Domaine Pero Longo – Lieu dit Navara – 20100 Sartene
Tel/Fax: 04 95 77 10 74
06 71 62 44 10
E-mail: perolongo@aol.com (en biodynamie)

APPENDIX | 157

Jura

Stéphane TISSOT – Domaine
André et Mireille Tissot –
39600 Montigny les Arsures
Tel: 03 84 66 08 27
Fax: 03 84 66 25 08
E-mail:stephane.tissot.arbois@
wanadoo.fr (en biologie)

***

Jean Etienne PIGNIER – Domaine Pignier – Cellier des
Chartreux – 39570 Montaigu
Tel: 03 84 24 24 30
Fax: 03 84 47 46 00
E-mail: pignier-vignerons@
wanadoo.fr (en biodynamie)

Languedoc – Roussillion

Cyril FAHL – Clos du Rouge
Gorge – 6 place Marcel Vie –
66720 Latour de France
Tel/Fax: 04 68 29 16 37
E-mail: cyrilfhal@tele2.fr (en
biodynamie)

***

Didier BARRAL – Domaine
Léon Barral – Lenthéric –
34480 Cabrerolles
Tel: 04 67 90 29 13
Fax: 04 67 90 13 37 (en biologie)

***

Jean François DEU – Domaine
du Traginer – 56 Av. Du Puig sur
Mer – 66650 Banyuls sur Mer
Tel: 04 68 88 15 11
Fax: 04 68 88 31 48
E-mail: jfdeu@hotmail.com
(en biodynamie)

***

Christophe BEAU – Domaine
Beauthorey – 30260 Corconne
Tel: 04 66 77 13 11
Fax: 04 66 77 12 06
E-mail: beau.corconne@wanadoo.fr (en biodynamie)

***

Bernard BELLHASEN – Domaine de Fontedicto – Fontareche – 34720 Caux

Tel: 04 67 98 40 22
Fax: 04 67 98 40 22
E-mail:fontedicto@tele2.fr (en
biodynamie)

***

AnneMarieLAVAYSSE–Lepetit
Domaine de Gimios – 34360
St Jean de Minervois
Tel/Fax: 04 67 38 26 10 (en
biodynamie)

***

Emmanuel CAZES – Domaine
Cazes – 4 rue Fransisco ferrer –
66600 Rivesaltes
Tel: 04 68 64 08 26
Fax: 04 68 64 69 79
E-mail: emmanuel.cazes@
cazes.com (en biodynamie)

Loire

Matthieu BOUCHET – Domaine de Château Gaillard
– Ruette du Moulin – 49260
Montreuil Bellay
Tel: 02 41 52 31 11
Fax: 02 41 52 39 94
E-mail: bouchet.matthieu@
wanadoo.fr (en biodynamie)

***

François PLOUZEAU – Domaine de la Garrelière – 37120
Razines
Tel: 02 47 95 62 84
Fax: 02 47 95 67 17
E-mail: francois.plouzeau@
wanadoo.fr

***

Marc ANGELI – Ferme de la
Sansonnière – 49380 Thouarcé
Tel/Fax: 02 41 54 08 08 (en
biodynamie)

***

Michel AUGE – Domaine des
Maisons Brulées – 5 Impasse de
la Vallée du Loin – 41110 Pouillé
Tel: 02 54 71 58 23
Fax: 02 54 71 51 57 (en biodynamie)

***

Philippe et Françoise GOURDON – Château Tour Grise –

1 rue des Ducs d'Aquitaine –
49260 Le Puy Notre Dame
Tel: 02 41 38 82 42
Fax: 02 41 52 39 96
E-mail: philippe.gourdon@
latourgrise.com (en biodynamie)

***

Nicolas JOLY – Clos de la
Coulée de Serrant – Château
de la Roche aux Moines –
49170 Savennières
Tel: 02 41 72 22 32
Fax: 02 41 72 28 68
E-mail: coulee-de-serrant@
wanadoo.fr (en biodynamie)

***

Guy BOSSARD – Domaine de
l'Ecu – La Bretonnière – 44430
Le Landreau
Tel: 02 40 06 40 91
Fax: 02 40 06 46 79
E-mail:bossard.guy.muscadet@
wanadoo.fr (en biodynamie)

***

Thierry MICHON – Domaine
Saint Nicolas – 11 rue des
Valées – 85340 Brem sur Mer
Tel: 02 51 90 55 74/02 51 33
13 04
Fax: 02 51 33 18 42
E-mail: contact@domainesaint-nicolas.com (en biodynamie)

***

Pierre BRETON – Les Galichets – 8 rue du Peu Muleau –
37140 Restigné
Tel: 02 47 97 30 41
Fax: 02 47 97 46 49
E-mail: catherineetpierre.
breton@libertysurf.fr (en biodynamie)

***

Joel MENARD – Domaine des
Sablonnettes – L'Espérance –
49750 Rablay sur Layon
Tel: 02 41 78 40 49
Fax: 02 41 78 61 15
E-mail: domainedessabl
onnettes@wanadoo.fr (en biodynamie)

***

René MOSSE – Domaine Mosse – 4 rue de la Chauvière – 49750 St Lambert du Lattay
Tel: 02 41 66 52 88
Fax: 02 41 68 22 10
Mob: 06 84 08 89 06
E-mail: domaine.mosse@wanadoo.fr (en biodynamie)

***

Olivier COUSIN – 17 rue Fontencau – 49540 Martigné Briand
Tel: 02 41 59 49 09 / 02 41 59 68 44
Fax: 02 41 59 69 83
Email: ocousinvin@wanadoo.fr (en biodynamie)

Provence

Yves GROS – Domaine les Fouques SCEA – 1405 Route des Borrels – 83400 Hyeres
Tel: 04 94 65 68 19
Fax: 04 94 35 25 30
E-mail: fouquesbio@wanadoo.fr (en biodyanmie)

***

Raimond de VILLENEUVE – Château de Roquefort -13830 Roquefort la Bédoule
Tel: 04 42 73 20 84
Fax: 04 42 73 11 19
E-mail: chateau.de.roquefort@free.fr (en biodynamie)

***

Château Romanin – 13210 Saint Rémy de Provence
Tel: 04 90 92 45 87
Fax: 04 90 92 24 36
E-mail: contact@romanin.com (en biodynamie)

***

Dominique HAUVETTE – Domaine Hauvette – La Haute Galine – 13210 St Remy de Provence
Tel: 04 90 92 03 90
Fax: 04 90 92 08 91
E-mail: domainehauvette@wanadoo.fr (en biodynamie)

***

Eloi DÜRRBACH – Domaine de Trevallon – 13103 Saint Etienne du Gres

Tel: 04 90 49 06 00
fax: 04 90 49 02 17
E-mail: trevallon@wanadoo.fr (en biologie)

***

Anne DUTHEIL de la RO-CHERE – Château Sainte Anne – Sainte Anne D'evenos – 83330 Evenos
Tel: 04 94 90 35 40 6
Fax: 04 94 90 34 20
E-mail: chateausteanne@free.fr (en biologie)

Rhône

Matthieu BARRET – Domaine du Coulet – 43 rue du Ruisseau – 07130 Cornas
Tel/Fax: 04 75 80 08 25
E-mail: domaineducoulet@tele2.fr (en biodynamie)

***

Marc GUILLEMOT – SCEA de Quintaine – Quitaine – 71260 Clesé
Tel: 03 85 36 95 88
Fax: 03 85 36 91 50
E-mail: sceadequitaine@wanadoo.fr (en biodynamie)

***

Dany et Fernand CHASTAN – Clos du Joncuas – 84190 Gigondas
Tel: 04 90 65 86 86
Fax: 04 90 65 83 68
E-mail: clos-du-joncuas@wanadoo.fr (en biologie)

***

Daniel BOULE – Domaine Les Aphillantes – EARL Les Galets – Quartier St Jean – 84850 Travaillan
Tel/Fax: 04 90 37 25 99
E-mail: lesgalet84@wanadoo.fr (en biodynamie)

***

Jacqueline ANDRE – Domaine Pierre André – 30 Faubourg Saint Georges – 84350 Courthezon
E-mail: domaine.pierre.andre@wanadoo.fr (en biodynamie)

***

Maxime GERVAIS – Domaine de Villeneuve – Route de Courthézon – 84100 Orange
Tel: 04 90 34 57 55
Fax: 04 90 51 61 22
E-mail: divinia@wanadoo.fr (en biodynamie)

***

Christine et Eric SAUREL – Montirius – Le Devès – 84260 SARRIANS
Tel: 04 90 65 38 28
Fax: 04 90 65 48 72
E-mail: montirius@wanadoo.fr (en biodynamie)

***

Alain et Philippe VIRET – Domaine Viret – Quartier les Escoulenches – 26110 St Maurice/Eygues
Tel: 4 75 27 62 77
Fax: 04 75 27 62 31
E-mail:cosmoculture@domaine.viret.com (en biodynamie)

***

Michel CHAPOUTIER – Maison Chapoutier – 18 avenue du Docteur Paul Durand – 26600 Tain L'hermitage
Tel: 04 75 08 28 65
Fax: 04 75 08 81 70
E-mail: cchapoutier@chapoutier.com (en biodynamie)

Savoy

Michel et Roselyne GRISARD – Domaine Prieuré Saint Christophe – 73250 Fréterive
Tel: 04 79 28 62 10
Fax: 04 79 28 61 74
E-mail: michelgrisard@cario.fr (en biodynamie)

South West

Yvonne HEGOBURU – Domaine de Souch – 64110 Jurançon
Tel: 05 59 06 27 22
Fax: 05 59 06 51 55
E-mail: jr.hegoburu@wanadoo.fr (en biodynamie)

***

Floreal ROMERO – Domaine
Le Bouscas – 32330 Gondrin
Tel: 05.62.29.11.87
Fax: 05.62.29.15.82
E-mail: romero-floreal@wana
doo.fr (en biodynamie)

Switzerland

Jaques GRANGE – Domaine
de Beudon – 1926 Fully Valais
SUISSE
Tel/Fax: 0041 27 744 1275
(en biodynamie)

Germany

Rainer EYMANN – Weingut
Eymann – Ludwigstrasse 35 –
D-67161 Gönnheim
Tel: 0049 63 22 28 08
Fax: 0049 63 22 687 92
E-mail: info@weinguteymann.
de (en biologie)

***

Peter et Martina LINXWEIL-
ER – Weingut Hahnmülle –
D-67822 Mannweiler-Cöln
Tel: 0049 63 62 99 30 99
Fax: 0049 63 62 44 66
E-mail: info@weingut-hahnm
uehle.de (en biodynamie)

***

PhilippWITTMANN – Wein-
gut Wittmann -19 Mainzer
Srtasse – D-67593 Westhofen
bei Worms
Tel: 0049 62 44 90 50 36
Fax: 0049 62 44 55 78
E-mail:info@weingut-wittmann.
de (en biodynamie)

***

StefanSANDER–WeingutSan-
der – In den Weingarten11 –
D-67582 Mettenheim
Tel: 0049 62 42 15 83
Fax: 0049 62 42 65 89
E-mail: info@weingut-sander.
de (en biologie)

Austria

Christine SAAHS – Nikolaihof
Wachau – A-3512 Mautern

Tel: 0043 27 32 82 901
Fax: 0043 27 32 76 440
E-mail: wein@nikolaihof.at
(en biodynamie)

***

Günther SCHÖNBERGER –
Weingut Schönberger – Setz-
gasse 9 – A-7072 Mörbisch
Am See
Tel: 0043 31 19 28 42
Fax: 0043 31 19 28 422
E-mail: ente@weingut-sch-
oenberger.com
(en biodynamie)

***

Ilse MAIER – Weingut Gey-
erhof – Oberfucha 1 – 3511
Furth
Tel: 0043 27 39 22 59
Fax: 0043 27 39 22 594
E-mail: weingut@geyerhof.at
(en biologie)

Slovenia

Ales KRISTANCIC – Movia –
Ceglo 18 – 5212 Dobrovo
Tel: 00 386 5 395 95 10
Fax: 00 386 5 395 95 11
E-mail: movia@siol.net (en
biodynamie)

Spain

Josep Maria ALBET I
NOYA – Albet I Noya – Can
Vendrell de la Codina –
08739 San Paul d'Ordal
Barcelona
Tel: 0034 938 99 48 12
Fax: 0034 938 99 49 30
E-mail: josepmaria@albetin
oya.com (en biologie)

***

Telmo RODRIGUEZ (contact
Conchi Garcia) – Compania
de vinons Telmo Ridriguez
Siete Infantes de Lara 5 – Ofic-
ina 1
26006 Logrono
Tel: 0034 941 511 128
Fax: 0034 941 511 131
E-mail: cia@fer.es (en biody-
namie)

***

Ricardo Pérez PALACIOS –
Descendientes de J. Palacios,
S.L – Calle Calvo Sotelo 6
24500 Villafranca del Bierzo
Tel: 0034 987 54 08 21
Fax: 0034 987 54 08 51
E-mail: djpalacios@mail.
ddnet.es (en biodynamie)

***

Alvaro PALACIOS – Alvaro Pal-
acios, S.L – 43737 Gratallops
Tel: 0034 977 83 91 95
Fax: 0034 977 83 91 97
E-mail: alvaropalacios@ctv.es
(en biologie)

***

Peter SISSECK – Domino de
Pingus – Apto. de Correos 93 –
47300 Penafiel (Valladolid)
Tel/Fax: 0034 983 484 002
E-mail: ps@pingus.es (en bio-
dynamie)

***

Diego SOTO – Mas Estela –
17489 Selva de Mar
Tel: 0034 972 12 61 76
Fax: 0034 972 38 80 11
E-mail: masestela@hotmail.
com (en biodynamie)

***

Edorta Lezaun Etxalar
E-mail: info@lezaun.com

***

Quinta Sardonia
E-mail: jbgnd@telefonica.net

Italy

Nadia RIGUCCINI – Az. Agri-
cola Campi Nuovi – Via Petro
Nenni 29 – 53011 Castellina
in Chianti (SI)
Tel: 0577 74209
E-mail: info@campinuovi.com

***

Mattia MAZZANTI – Az Agr.
Podere la Cerreta – Via Campgna
sud 143 – 57020 Sasseta (LI)
Tel: 0565 794352
E-mail: cerreta@cerreta.it

***

Silvio MESSANA – Montesec-
ondo – Via per Cerbaia 18 –
50020 Cerbaia val di Pesa

Tel: +39 055 825 208
Fax: +39 055 825 98 28
E-mail: montesecondo@
montesecondo.com

***

Gabriele BUONDONNO – Loc. Casavecchia a la piazza, 37 – 53011 Castellina in Chianti (SI)
Tel: +39 0577749754
Fax +39 0577733662
E-mail: buondonno@chianti classico.com (en biologie)

***

Umberto VALLE – Poggio Tewalle S.S – Podere ex Ente Maremma 348 – Loc. Arcille 58050 Campagnatico (GR) – Toscana – Italia
Tel/Fax 0039 564 998142
E-mail: valle@poggiotrevvalle. it (en biologie)

***

Stella DI CAMPALTO – Az. Agr. San Giusepe – Podere San Giuseppe 35 –
CAP 53024 Castelnuovo dell'Abate – Montalcino (SI)
Tel/Fax: +39 0577 835754
E-mail: stella.violadicampalto@ tin.it (en biodynamie)

***

Fabio MONTOMOLI – FATTORIA CASTELLINA – Via Palandri, 27 50050 Capraia e Limite (FI) – IT/F +39 0571 57631
E-mail: info@fattoriacastel lina.com (en biodynamie)

***

Antoine LUGINBÜHLl – Casina di Cornia – Località Cornia 113 – 53011 Castellina in Chianti (Siena) Phone: +39-0577-743053
Fax: +39-0577-743059
E-mail: info@casinadicornia. com (en biologie)

***

Stefano BELLOTTI – Casina Degli Ulivi – Strada Mazzola 14 – 15067 Novi Ligure (AL)

Tel: 0039 01 43 74 45 98
Fax: 0039 01 43 32 08 98
E-mail: cascinadegliulivi@ libero.it (en biodynamie)

***

Andrea SALVETTI – Cascina La Pertica – Via Rosario 44 – 25080 Polpenazze del Garda (Brescia)
Tel: +39 03 65 65 14 71
Fax: +39 03 65 65 19 91
E-mail: asalvetti@cascinalap ertica.it (en biodynamie)

***

Massimiliano Baldini LIBRI – Fattoria Cerreto Libri – Via Aretina 90 – 50065 Pontassieve (FI)
Tel: 0039 055 8314 528
E-mail: fattoria@cerretolibri.it (en biodynamie)

***

Laura di COLLOBIANO/ Moreno PETRINI – Tenuta di Valgiano – via di Valgiano 7, 55018 Valgiano – Lucca
Tel: +39 0583 402271
E-mail: saverio@valgiano.it (en biodynamie)

***

Denis MONTANAR – Domaine BORC DODON – V. Malborghetto, 4 – 33059 – Villa Vicentina (UD)
Tél: +39 (0)431 969393
E-mail: denismontanar@libero.it

***

Rainer LOACKER – Loacker Tenute – via Santa Justina 3 – 1 • 39100 BOZEN/BOLZANO T +39 0471 365125
E-mail: scf.lo@cker.it

***

Sofia PEPE – Az. Agricola Bio Emidio Pepe – Via Chiesi n. 10 – 64010 Torano Nuovo TE Abruzzo Italy
E-mail: info@emidiopepe.com

***

Rosela BECINI TESI – Terre a Mano – Via Fontemorana 179 – 59015 Bacchereto – Carmignamo (Po)
Tel/Fax: 0039 055 87 17 191
E-mail: fattoriadibacchereto@ libero.it (en biodynamie)

***

Giuseppe FERRUA – Fabbrica di San Martino – Via Pieve Santo Stefano 2511 – 55100 Lucca (Lu)
Tel/Fax: 0039 0583 39 42 84
E-mail: info@fabbricadisan martino.it (en biodynamie)

***

Paolo FRANCESCONI – Via Tuliero 154 – 48018 Faenza (Ra)
Tel: 0039 0546 43 213
E-mail: pfrancesconi@racine. ra.it

***

Florio GUERRINI – Il Paradiso di Manfredi – Via Canalicchio 305 – 53024 Montalcino (Si)
Tel: 0039 0577 84 84 78
E-mail: ilparadisomanfredi@ interfree.it

***

Emilio FALCIONE – La Busattina – Loc Busattina, 58050 San Martino sul Fiora (GR)
Tel/Fax: 0039 0564 60 78 40
Email: busattina@tiscali.it

***

Podere CONCORI – Via Provinciale 1 Fiattone – 55027 Gallicano Lucca,
Tel/Fax: 0583766374
Email: gabdapr@tin.it

***

AZ Agr. San Polino – Pod. San Polino 163 – 53024 6 Montalcino
Tel: 0577 835775
E-mail: vino@sanpolino.it

***

Fabrizio ROSSI – Az Agr. Cefalicchio – Corso San Sabino 6 – 70053 Canosa di Puglia (BA)
Tel: 08883 617601
Fax: 0883 666238
E-mail: info@cefalicchio.com

United States

Mike and Chris BENZIGER – Benziger Family Winery – 1883 London Ranch Road,

Glen Ellen – CA 95442
Tel: 001 707 935 4066
Fax: 001 707 935 3018
E-mail: chrisbz@benziger.
com
(en biodynamie)

***

Robert SINSKEY – Robert Sinskey Vineyard – 6320 Silverado trail – Napa Califonia, 94558
Tel: 001 707 944 90 90
E-mail: rsv@robertsinskey.com
(en biologie)

***

Doug TUNNEL & Melissa MILLS – Brich House Wine Compagny – 18200 Lewis Rogers Lane, Newberg Oregon 97132
Tel/Fax: 001 503 538 1 36
E-mail: doug@brickhousewines.
com (en biodynamie)

***

Bart & Daphne ARAUJO – Araujo Estate Wines
Tel: 001 707 942 6061
Fax: 001 707 942 64 71
E-mail: wine@araujoestate.com

***

John WILLIAMS (contact Jonah Beer)– Frogs Leap Winery – 8815 Conn Creek Rd. – Rutherford CA 94573
Tel: 001 707 963 4704
Fax: 001 707 963 0742
E-mail: rabbit@frogsleap.com

***

Tony COTTURRI – Cotturri Winery – 6725 Enterprise rd – Glen Ellen – CA 954442-9504
Tel: 001 707 525 9126
Fax: 001 707 542 8039
E-mail: tony@coturriwinery.com

***

Jim FETZER – Ceago Vinegarden – PO Box 3017 – 5115 East Hwy 20 Nice – CA 95464
Tel: 707 485 5799
Fax: 707 485 6061
E-mail: ceago@ceago.com

***

Robert BLUE – Bonterra Vineyards – 2231 McNab Ranch Road – Ukiah – CA 95482
Tel: 707 462 7814
E-mail: bob_blue@B-F.com

***

Patti FETZER BURKE – Patianna – 6552 Red Wineyard Road – Geyserville – CA 95441
E-mail: patti@patianna.com

***

Russel RANEY – Evasham Wood Winery – 3795 Wallace Rd NW – Salem Oregon 97304
Tel: 503 371 8478
Fax: 503 763 6015
E-mail: evewood@open.org (en biologie)

***

Christophe BARON – Cayuse Vineyard – PO box 1602 – Walla Walla WA 99362
Tel: 001 509 526 0686
Fax: 001 509 526 4686
E-mail: info@cayusevineyards.com

***

Julie Johnson – Tres Sabores – 620 S. Whitehall Lane – St Helena – CA 94574
Tel: 001 707 967 8027
E-mail: jaj@tressabores.com

***

Stephen Ryan – Mendocino Farms – 6201 Old River Road – Ukiah, CA 95482
Tel: 707-462-3688
Fax: 707-462-7246
E-mail: steve@mendofarms.com

South Africa

Johan REYNEKE – Reyneke Wines – Uitzicht farm – P.O. box 61 – Viottenburg 7604
Tel: 0027 21 881 35 17
Fax: 0027 21 881 35 17
E-mail: wines@reynekewines.co.za (en biologie)

***

Kurt AMMANN – Rozendal farm – Omega Street – Jonkershoek Valley – Stellenbosch 7600
Email: rozendal@mweb.co.za

New Zealand

James & Annie MILLTON – The Millton Vineyards – Papatu Rd Manutuke Po Box 66 Manutuke Poverty Bay
Tel: 0064 6 862 86 80
Fax: 0064 6 862 88 69
E-mail: james@millton.co.nz
(en biodynamie)

Australia

Julian & Carolann CASTAGNA – Castagna Vineyard – 88 Ressom Lane – PO Box 73 – Beechworth Victoria 3747
Tel: 0061 3 57 28 28 88
Fax: 0061 3 57 28 28 98
E-mail: castagna@enigma.com.
au (en biodynamie)

***

Ron LAUGHTON – Jasper Hill Vineyard – Drummonds Lane – PO Box 110 – Heathcote Victoria 3523
Tel: 0061 3 54 33 25 28
Fax: 0061 3 54 33 31 43
E-mail: ronlaughton@ozemail.
com.au (en biologie)

***

Vanya CULLEN – Cullen Wines – Pty Ltd Box 17 Cowaramup 6284 Western Australia
Tel: 08 9755 5277
Fax: 08 9755 5550
Email: enquiries@cullenwines.
com.au (en biologie)

South America

Alvaro ESPINOZA (contact Marina Ashton) – Vina Antiyal – PO Box 191, Paine – CHILI
Tel: 0056 2 821 42 24
E-mail: marina@antiyal.com
(en biodynamie)

***

Rafael GUILISASTI (contact Fransisca Marshall) – Vinedos santa Emiliana (VOE) – Av, Nueva Tajarnar 481 – Of 401 – Torre Sur, Las Coondes – Santiago – CHILI
Tel: 00 562 353 91 30
Fax: 00 562 203 62 27
E-mail: ijofre@emiliana.cl (en biodynamie)

Demeter Certified & In-Conversion Biodynamic Wineries, Vineyards & Farms U.S., 2007

Certified Wineries, Vineyards & Farms

AmByth Estate
2314 Westminster Ave.
Costa Mesa
CA 92627
805-305-9497
Wine grapes: Tempranillo, Syrah, Voignier, Mourvedre, Grenache, Sangiovese; olives

Araujo Estate - Holopono Vineyard
2155 Pickett Rd.
Calistoga
CA 94515
707-942-6061
www.araujoestate.com
Wine grapes: Cabernet Sauvignon, Syrah, Sauvignon Blanc, Sauvignon Musque, Voignier, Cabernet Franc, Petit Verdot, Merlot; olives

Benziger Family Vineyards
1883 London Ranch Rd.
Glen Ellen
CA 95442
707-935-4066
www.benziger.com
Wine grapes: Cabernet Sauvignon, Merlot, Cabernet Franc, Petit Verdot, Zinfandel, Sauvignon Blanc, Malbec, Pinot Noir, Chardonnay, Cabernet, olives, mixed fruits & vegetables

Benziger Family Winery
1883 London Ranch Rd.
Glen Ellen
CA 95442
707-935-4066

www.benziger.com
Wines; **Made From Biodynamic Grapes** 2001 Sonoma Mountain Red, 2001 Tribute, 2002 Sonoma Mountain Red, 2002 Tribute, 2003 Sonoma Mountain Red, 2003 Sonoma Mountain Zinfandel, 2003 Sonoma Mountain Sauvignon Blanc, 2003 Tribute, 2004 Sonoma Mountain Red, 2004

Bergstrom Vineyard
18405 NE Calkins Ln.
Newberg
OR 97132
503-554-0468
www.bergstromwines.com
Wine grapes: Pinot Noir

Bergstrom Winery
18405 NE Calkins Ln.
Newberg
OR 97132
503-554-0468
www.bergstromwines.com
Wine; **Biodynamic Wine;** 2004 Bergstrom Pinot Noir, 2005 deLancellotti Pinot Noir

Bonny Doon Ca 'del Solo Vineyard
29748 Camphora-Gloria Rd.
Soledad
CA 93960
831-235-2313
Wine grapes; Dolcetto, Albarino, Freisa, Riesling, Syrah, Loureino, Grenache, Pinot Noir, Orange Muscat, Linsault, Grenache Blanc, Muscat Giallo

Bonny Doon Winery
328 Ingalls Street
Santa Cruz
CA 95060
831-425-6763
Wine

Bonterra Vineyards at McNab Ranch
2231 McNab Ranch Rd.
Ukiah
CA 95482
707-272-4655
Wine Grapes

Bonterra Vineyards at Butler Ranch
5500 Butler Ranch Rd.
Ukiah
CA 95482
707-272-4655
www.bonterra.com
Wine grapes: Syrah, Zinfandel, Cabernet, Malbec, Grenache, Petit Verdot, Mouvedre; cherries

Brickhouse Winery
18200 Lewis Rogers Rd.
Newberg
OR 97132
503-538-5136
Wine; **Made From Biodynamic Grapes;** 2005 Les Dijonnais Pinot Noir, 2005 Select Pinot Noir, 2005 Evelyns Pinot Noir, 2005 Chardonnay. **Biodynamic Wine;** 2005 Gamay Noir

Brickhouse Vineyard
18200 Lewis Rogers Rd.
Newberg
OR 97132
503-538-5136
Wine grapes: Pinot Noir, Gamay Noir, Chardonnay;

Cayuse Vineyards
PO Box 1602
Walla Walla
WA 99362
509-526-0686
www.cayusevineyards.com
Wine grapes: Syrah, Grenache, Mouvedre, Voignier, Cabernet Sauvignon, Merlot, Tempranillo, Cabernet Franc

Ceago Vinegarden--del Lago Ranch
5115 Highway 20 East
Nice
CA 95464
707-274-1462
Wine grapes: Malbec, Sirah, Petit Sirah, Chardonnay, Muscat, Cabernet, Sauvignon Blanc, Semillon, Gewurstraminer, Petite Verdot, Cabernet Franc; Walnuts, lavender, olives

Ceago Vinegarden Winery
1160 Bel Arbes Rd.
Redwood Valley
CA 95470
707-274-1462
Wines; **Made From Biodynamic Grapes**; 2004 Camp Masut Merlot, 2005 Del Lago Muscat Canelli, 2005 Del Lago Syrah Rose', 2005 Kathleens Sauvignon Blanc, 2005 Del Lago Cabernet Franc, 2005 Del Lago Malbec, 2005 Del Lago Syrah, 2005 Clear Lake Winemakers Blend, 2

Chequera Vineyard
2485 Hwy. 46 West
Paso Robles
CA 93446
559-246-8426
Wine Grapes

Cooper Mountain Vineyards & Winery
9480 Grabhorn Rd.
Beaverton
OR 97007
503-649-0702
Wine grapes: Pinot Noir, Pinot Gris, Chardonnay, Pinot Blanc; Tokai (In-Transition);

Cooper Mountain Winery
9480 Grabhorn Rd.
Beaverton
OR 97007
503-649-0702
Wine; **Biodynamic Wine**; 2005 Meadowlark Pinot Noir, 2005 Mountain Terrior Pinot Noir, 2005 Old Vines Chardonnay, 2005 Old Vines Pinot Noir, 2006 Cooper Hill Pinot Gris, 2006 Harmony Blend, 2006 Life Pinot Noir, 2006 Old Vines Pinot gris, 2006 Reserve Chard

Cow Horn Vineyard
1665 Eastside Rd.
Jacksonville
Or 97530
541-301-7038
Wine grapes; Syrah, Viognier, Grenache, Rousanne, Marsanne. Asparagus

Cow Horn Winery
1665 Eastside Rd.
Jacksonville
OR 97530
541-301-7038
Biodynamic Wine; 2006 Cow Horn Viognier, Made From Biodynamic Grape Wine; 2006 Cow Horn Marsanne

Dark Horse Vineyard
534 Old River Rd.
Ukiah
CA 95482
707-391-7554
Wine Grapes: Cabernet, Zinfandel, Petit Sirah, Syrah, Grenache; olives

Fetzer Vineyards Winery
PO Box 611
Hopland
CA 95449
707-744-1521
Wine, **Made With Biodynamic Grapes**; 2002 McNab Ranch Blend, 2003 Mcnab Ranch Blend

Frey Vineyards
14000 Tomki Rd.
Redwood Valley
CA 95470
707-485-5177
www.freywine.com
Wine grapes: Cabernet Sauvignon, Zinfandel, Petit Sirah, Merlot, Chardonnay, Sauvignon Blanc, Syrah, Sangiovese; sage, oregano, rose, Douglas fir, rosemary, apples

Frey Winery
14000 Tomki Rd.
Redwood Valley
CA 95470
707-485-5177
www.freywine.com
Biodynamic Wine; 2005 Sauvignon Blanc, 2005 Syrah, 2005 Zinfandel, 2005 Cabernet Sauvignon, 2005 Petite Sirah, 2005 Merlot, 2005 Chardonnay 2004 Merlot, 2004 Cabernet Sauvignon, 2004 Zinfandel, 2004 Sauvignon Blanc, 2004 Petite Sirah, 2003 Pinot Noir Masut

Golden Vineyard--Heart Arrow
Heart Arrow Trail
Hopland
CA 95449
707-485-8885
Wine Grapes: Petit Sirah, Cabernet Sauvignon, Zinfandel: olives

Golden Vineyard--Fairbairn Vineyard
14201 Old River Road

Hopland
CA 95449
707-485-8885
Wine grapes: Syrah

Grgich Hills Vineyards
PO Box 450, 829 St Helena
Hwy
Rutherford
CA 94573
707-963-2784
Wine grapes; Sauvignon Blanc,
Chardonnay, Gewurztraminer,
Merlot, Reisling

Grgich Hills Winery
PO Box 450, 1829 St Helena
Hwy
Rutherford
CA 94573
707-963-2784
Wine; Made From Biodynamic
Grapes 2006 Essence Sauvignon Blanc

Harms Vineyard & Lavender
Fields
3185 Dry Creek Rd.
Napa
CA 94558
707-257-2602
www.harmsvineyardsandlav
enderfields.com
Lavender, rosemary, lemon
verbena; wine grapes: Chardonnay, Sangiovese; lavender
oil, hydrosol

La Clarine Farm
PO Box 245
Somerset
CA 95684
530-626-6964
Wine grapes: Syrah, Grenache,
Tempranillo

Marian Farms Distillery
POB 9167
Fresno
CA 93790
559-276-6185
Brandy

Masut Vineyard
POB 348
Redwood Valley
CA 95470
707-485-5466
www.masut.com
Wine grapes: Sangiovese,
Pinot Noir

Mendocino Farms Winery
6200 Old River Rd.
Ukiah
CA 95482
707-953-1044
Wine; Made From Biodynamic
Grapes 2004 Red Vine Series,
2005 Syrah Fairburn Ranch

Mendocino Wine Group
501 Parducci Rd.
Ukiah
CA 95482
707-463-5377
Wine; Made From Biodynamic
Grapes; 2005 Mendocino
Deep Red

Momtazi Vineyard
15765 Muddy Valley Road
McMinnville
OR 97128
503-843-1234
Wine grapes: Pinot Noir, Pinot Gris, Chardonnay

Patianna Organic Vineyards
10291 East Side Rd.
Ukiah
CA 95482
707-433-4097
www.patianna.com
Wine grapes: Sauvignon
Blanc, Chardonnay

Patianna Vineyards - wine
10291 East Side Rd.
Ukiah
CA 95482
707-433-4097
www.patianna.com
Wines: Made From Biodynamic Grapes 2005 Mendo-

cino Syrah, 2006 Mendocino
Sauvignon Blanc

Porter Creek Vineyards
8735 Westside Road
Healdsburg
CA 95448
707-433-6321
www.portercreekvineyards.
com
Wine grapes: Pinot Noir,
Chardonnay

Presidio Vineyards
2755 Purisima Rd.
Lompac
CA 93436
805-740-9463
www.presidiowinery.com
Wine grapes: Pinot Noir, Syrah, Chardonnay, Viognier

Quivira Vineyards
4900 West Dy Creek Road
Healdsburg
CA 95448
707-431-1664
www.quivirawine.com
Wine grapes: Grenache Noir,
Sauvignon Blanc, Sauvignon
Musque, Semillon, Zinfandel,
Mourvedre, Viognier, Petite
Sirah, Syrah

Quivira Winery
4900 West Dy Creek Road
Healdsburg
CA 95448
707-431-1664
www.quivirawine.com
Wine

Resonance Vineyard
12000 NW Foothills Rd.
Carlton
OR 97111
503-852-6373
Wine grapes: Pinot Noir,
Gewurztraminer

Resonance Winery
12000 NW Foothills Rd.
Carlton
OR 97111

503-852-6373
Wine

Robert Sinskey Vineyard
6320 Silverado Trail
Napa
CA 94558
707-251-0469
Wine grapes: Pinot Noir, Gewurztraminer, Reisling,

Muscat, Pinot Gris, Pinot Blanc

Rose Ranch
PO Box 1253
Kenwood
CA 95452
707-833-2143
Wine grapes; Merlot, Primativo. Mixed vegetables

Sycamore Vineyard
600 Montgomery St 35th floor
San Francisico
CA 94111
415-421-9990
Wine grapes; Cabernet, Cabernet Franc, Merlot

In-Conversion Wineries, Vineyards & Farms

Beasley Vineyard
964 Chiquita Rd.
Healdsburg
CA 95448
707-431-7985
Zinfandel wine grapes

Enterprise Vineyards
PO Box 233
Vineburg
CA 95487
707-996-6513
Wine grapes

Honey Creek Dairy Farm
W2716 Friemoth Road
East Troy
WI 53120
262-642-2207
Dairy, Milk, Hay

Jack Rabbit Hill Vineyards & Winery
IC
Anna & Lance Hanson
PO Box 2004 , 26567 North Road
Hotchkiss
CO 81419
970-835-3677
Wine Grapes, Wine

Leopardo Olive Grove
1478 Olivet Rd.
Santa Rosa
CA 95401

707-526-3177
Olives

By The Light Of Day Farm
13191 S.Partridge Run Dr
Traverse City
MI 49684
231-228-7234
Herbs for tea blends

Montinore Estate
PO Box 490
Forest Grove
OR 97116
503-359-5012
Wine Grapes, Wine

Mt. Hood Organics
7130 Smullin Road
Mt. Hood
OR 97041
541-352-7492
Apples, pears

Nezinscot Farm
284 Turner Center Rd.
Turner
ME 4282
207-225-3231
Dairy, veggies, meat, cheese

Ola Honua
HRC1 Box180
Hana

HI 96713
808-248-7561
Vegetables, bamboo forest

Old Fashioned Farm
25413 Pink Schoolhouse Road
Evans Mills
NY 13637
315-629-4409
Fruit, vegetables, dairy herd, beef herd, fluid milk, hay, pasture

Oregon's Wild Harvest
43464 Phelps Rd.
Sandy
OR 97055
503-668-7145
Medicinal herbs

Pacific Rim Winery
1111 East Burnside Ave.
Portland
OR 97211
831-901-7386
Wine

Peak Spirits
IC
Anna & Lance Hanson
PO Box 2004 , 26567 North Road
Hotchkiss
CO 81419
970-835-3677
Spirits

Puma Springs Vineyard
1421 Chiquita Road (physical) 1083 Vine Street Mailbox 286 (mailing)
Healdsburg
CA 95448
(707) 431-9173
Wine grapes

Quady Winery
PO Box 728
Madera
CA 93639
559.673.8068

Rebensdorf Vineyrad
4254 N. Jameson, 6199 N. Rolinda
Fresno
CA 93723
559-275-3710
Wine & table grapes

Resting in the River
PO Box 816
Abiquiu
NM 87510
505-685-4364

Rex Hill Vineyards - Pearl Vineyard
9315 NE Red Hills Rd.
Dundee
OR 97115
503-209-0431

Sequatchie Cove
320 Dixon Cove Rd.
Sequatchie
TN 37374
423-942-9201

Vegetables, fruit, dairy herd, hay

Skyline Organic Farms
PO Box 1156, Topanga Skyline Dr.
Topanga
CA 90290
310-455-1123
Fruit, herbs, wine grapes, nuts, eggs, chickens

Solar Living Institute
PO Box 836
Hopland
CA 95449
707-744-2017
Vegetables, fruit, herbs, nuts, hops

Sophias Garden
IC
Monique Camp
4038 Green Valley School Rd.
Sebastopol
CA 95472
707-823-2531
Apples, pears, plums

Sophia's Garden - processing
IC
Monique Camp
4038 Green Valley School Rd.
Sebastopol
CA 95472
707-823-2531
Apples, pears, plums

Sun Hawk Farms
2001 Duncan Springs Rd.
Hopland

CA 95449
707-972-9720

Grapes, Olives, vegetables
Sweet Earth Farm
43299 Patton Road
Gays Mills
WI 54631
608-872-2487

Thanksgiving Farm
PO Box 840
Harris
NY 12701
845-794-1400 x2257
Vegetables, cattle, swine

Three Swallows Farm
D/T
Jeff Marianni
23 Nelson Road
Ithaca
NY 14850
607-273-1046

Wallula Gap Vineyards
61603 N. Wilgus Rd.
Grandview
WA 98930
509-948-0284
Wine grapes

Weibel Winery
PO Box 443, 13300 Buckman Dr.
Hopland
CA 95449
707-744-2201
Wine

Bibliography and Further Reading

Adams, G. and Whicher, O., *The Plant Between Sun and Earth*, London 1980

Bott, V., *An Introduction to Anthroposophic Medicine*, Sussex 2004

Cloos, W., *The Living Earth*, Cornwall 1977

Cook, W.E., *The Biodynamic Food and Cookbook*, Sussex 2006

Cook, W.E., *Foodwise*, Sussex 2003

Edwards, L., *The Vortex of Life*, Edinburgh 1993

Grohmann, G., *The Plant, A Guide to Understanding Its Nature*, London 1974

Hauschka, R., *The Nature of Substance*, London 1983

Julius, Fritz H., *The Imagery of the Zodiac*, Edinburgh 1993

Kolisko, Drs. E. and L., *Agriculture of Tomorrow*, Bournemouth 1978

Pelikan, W., *Healing Plants*, New York 1997

Podolinsky, A., *Biodynamic Agriculture, Introductory Lectures*, vols. I and II, 1985

Pfeiffer, E., *Sensitive Crystallization Process*, New York 1975

Schwenk, T., *Sensitive Chaos*, London 1965

Soper, J., *Studying the 'Agriculture Course,'* West Midlands 1976

Steiner, R., *Agriculture Course*, Sussex 2004

Steiner, R., *The Spiritual Hierarchies*, New York 1996

Thun, M., *The Biodynamic Sowing and Planting Calendar*, Edinburgh 2006 (and yearly publication)

Thun, M.., *The Biodynamic Year*, Sussex 2007

Thun, M., *Gardening for Life*, Stroud 2000

Thun, M., *Results from the Biodynamic Sowing and Planting Calendar*, Edinburgh 2004

Contacts

Nicolas Joly – Clos de la Coulée de serrant – Château de la Roche aux
Moines – 49170 Savennières – France
Tel: 0033 (0)2 41 72 22 32 – Fax: 0033 (0)2 41 72 28 68
e-mail: coulee-de-serrant@wanadoo.fr
website: www.coulee-de-serrant.com

See the Demeter website for links to biodynamic organizations
around the world: www.demeter.net info@demeter.net

USA:
Biodynamic Farming and Gardening Association, Inc.
25844 Butler Road
Junction City
OR 97448
Tel.: 888 516-7797 or 541 998-0105
Fax: 541 998-0106
email: biodynamic@aol.com
www.biodynamics.com

UK

Biodynamic Agricultural Association

The Painswick Inn Project

Gloucester Street

Stroud

Glos. GL5 1QG

Tel./Fax: 01453 759501

email: office@biodynamic.org.uk

www.biodynamic.org.uk

Picture Credits

All photos and illustrations by Nicolas Joly, except for:

Cover: Robert Bruno

Page 7: Steiner, R., *Chaleur et Matiere*, Editions Triade, France

Plate 3: Les plantes médicinales, Reader's Digest

Page 58: *Book of Kells*, Trinity College; George Bain, celtic Art, Glasgo

Pages 63, 66, 69: Adam P., Wyss A., *Platonische und archimedische Körper, ihre Sternformen und polaren Gebilde*, Verlag freies Geistesleben

Page 65: Agrippa H.C., *De Philosophica Occulta, Book II, Of the proportions, measures and harmony of man's body* / Da Vinci L. *L'homme de Vitruve*

Page 67: Coates K., *Geometry Proportion and the Art of Luthery*, Oxford University Press 1985

Page 68: Ghyka, *Le nombre d'or*

Page 85, Plate 8: Snow Crystals, Dover Publications

Pages 87, 89: Bonvin J., Monterc R., *Elise Romane*, Edition Mosaique

Plate 11: by Bernard Morales

Plates 13, 14: Prieur B., Tesson MT., Association Presence (courtesy Nicolas Joly)

Plate 15: Christian Marcel (courtesy Nicolas Joly)

Pages 131, 140; Plate 16: Thun, M., *Sowing and Plantin Calendar*, Floris Books

Index

WINE BOOK PUBLISHER OF THE YEAR
Gourmand World Book Awards, 2004

The Wine Appreciation Guild has been an educational pioneer in our fascinating community.
—**Robert Mondavi**

Your opinion matters to us...

You may not think it, but customer input is important to the ultimate quality of any revised work or second edition. We invite and appreciate any comments you may have. And by registering your WAG book you are enrolled to receive prepublication discounts, special offers, or alerts to various wine events, only available to registered members.

Your first bonus for registering will be a free copy of our bestselling **GLOBAL ENCYCLOPEDIA OF WINE,** on CD ROM (a $29.95 value). This CD is compatible with PCs and Macs running Mac Classic. It has:

- Wine regions
- The process of grapes to glass
- Enjoying wine, rituals and tasting
- Wine Guide, a fascinating database for choosing different wines
- Cellar Log Book, that will allow you to document your own wine collection.

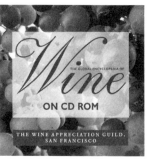

You can register your book by phone: (800) 231-9463; Fax: (650) 866-3513; E-Mail: Info@WineAppreciation.com; or snail mail the form on the following page.

REGISTRATION CARD

Name_____Date_____

Professional Affiliation_____

Address_____

_____

City_____State_____Zip_____

E-Mail_____

How did you discover this book?_____

Was this book required class reading? Y N

School/Organization_____

Where did you acquire this book?_____

Was it a good read? (circle)   Poor 1 2 3 Excellent

Was it useful to your work? (circle) Poor 1 2 3 Excellent

Suggestions_____

_____

Comments_____

_____

_____

You can register your book by phone: (800) 231-9463; Fax: (650) 866-3513; Email: Info@WineAppreciation.com; or snail mail.

THE WINE APPRECIATION GUILD
360 Swift Avenue
South San Francisco, CA 94080

www.wineappreciation.com

Fold Here ▲

Tape Closed Here ▼